FE MECHANICAL PRACTICE PROBLEMS

for the Mechanical Fundamentals of Engineering Exam

Michael R. Lindeburg, PE

Professional Publications, Inc. • Belmont, California

Report Errors and View Corrections for This Book

Everyone benefits when you report typos and other errata, comment on existing content, and suggest new content and other improvements. You will receive a response to your submission; other engineers will be able to update their books; and, PPI will be able to correct and reprint the content for future readers.

PPI provides two easy ways for you to make a contribution. If you are reviewing content on **feprep.com**, click on the "Report a Content Error" button in the Help and Support area of the footer. If you are using a printed copy of this book, go to **ppi2pass.com/errata**. To view confirmed errata, go to **ppi2pass.com/errata**.

Notice to Readers of the Digital Book
Digital books are not free books. All digital content, regardless of delivery method, is protected by the same copyright laws that protect the printed book. Access to digital content is limited to the original user/assignee and is non-transferable. PPI may, at its option, use undetectable methods to monitor ownership, access, and use of digital content, and may revoke access or pursue damages if user violates copyright law or PPI's end-use license agreement.

Personal Opinion Disclaimer
All views and opinions expressed in this book are those of Michael R. Lindeburg, PE in his individual capacity as the sole author of this book only. These views and opinions do not necessarily represent the views and opinions of Professional Publications, Inc., its employees, customers, consultants and contractors, or anyone else who did not participate in this book's authorship.

Disclaimer of Liability
This book should not be used for design purposes. With the exception of the explicit guarantee of satisfaction limited to the purchase price, this book is provided "as is" without warranty of any kind, either expressed or implied, unless otherwise stated. In no event shall the Author, Publisher, or their juridical persons be liable for any damages including, but not limited to, direct, indirect, special, incidental or consequential damages or other losses arising out of the use of or inability to use this book for any purpose. By keeping and using this book, readers agree to indemnify and hold harmless the Author and Publisher, its owners, employees, and its agents, from any damages claimed as a result of information, resources, products or services, or third party links obtained from this book and the PPI (**ppi2pass.com**) and FEPrep (**feprep.com**) websites.

FE Mechanical Practice Problems

Current printing of this edition: 3

Printing History

date	edition number	printing number	update
May 2014	1	1	New book.
Apr 2015	1	2	Minor corrections. Minor cover update.
Feb 2016	1	3	Minor corrections.

© 2014 Professional Publications, Inc. All rights reserved.

All content is copyrighted by Professional Publications, Inc. (PPI). No part, either text or image, may be used for any purpose other than personal use. Reproduction, modification, storage in a retrieval system or retransmission, in any form or by any means, electronic, mechanical, or otherwise, for reasons other than personal use, without prior written permission from the publisher is strictly prohibited. For written permission, contact PPI at permissions@ppi2pass.com.

Printed in the United States of America.

PPI
1250 Fifth Avenue
Belmont, CA 94002
(650) 593-9119
ppi2pass.com

ISBN: 978-1-59126-442-2

Library of Congress Control Number: 2014934952

F E D C B A

Topics

Topic I: Mathematics
Topic II: Probability and Statistics
Topic III: Fluid Mechanics
Topic IV: Thermodynamics
Topic V: Heat Transfer
Topic VI: Statics
Topic VII: Material Properties and Processing
Topic VIII: Mechanics of Materials
Topic IX: Electricity and Magnetism
Topic X: Dynamics, Kinematics, and Vibrations
Topic XI: Mechanical Design and Analysis
Topic XII: Measurement, Instrumentation, and Controls
Topic XIII: Computational Tools
Topic XIV: Engineering Economics
Topic XV: Ethics and Professional Practice

Where do I find help solving these Practice Problems?

FE Mechanical Practice Problems presents complete, step-by-step solutions for more than 470 problems to help you prepare for the Mechanical FE exam. You can find all the background information, including charts and tables of data, that you need to solve these problems in the *FE Mechanical Review Manual*.

The *FE Mechanical Review Manual* may be obtained from PPI at **ppi2pass.com** or **feprep.com**, or from your favorite print book retailer.

Table of Contents

Preface vii
Acknowledgments ix
How to Use This Book xi

Topic I: Mathematics

Analytic Geometry and Trigonometry 1-1
Algebra and Linear Algebra 2-1
Calculus 3-1
Differential Equations and Transforms 4-1
Numerical Methods 5-1

Topic II: Probability and Statistics

Probability and Statistics 6-1

Topic III: Fluid Mechanics

Fluid Properties 7-1
Fluid Statics 8-1
Fluid Dynamics 9-1
Fluid Measurement and Similitude 10-1
Compressible Fluid Dynamics 11-1
Fluid Machines 12-1

Topic IV: Thermodynamics

Properties of Substances 13-1
Laws of Thermodynamics 14-1
Power Cycles and Entropy 15-1
Mixtures of Gases, Vapors, and Liquids 16-1
Combustion 17-1
Heating, Ventilating, and
   Air Conditioning (HVAC) 18-1

Topic V: Heat Transfer

Conduction 19-1
Convection 20-1
Radiation 21-1

Topic VI: Statics

Systems of Forces and Moments 22-1
Trusses 23-1
Pulleys, Cables, and Friction 24-1
Centroids and Moments of Inertia 25-1

Topic VII: Material Properties and Processing

Material Properties and Testing 26-1
Engineering Materials 27-1
Manufacturing Processes 28-1

Topic VIII: Mechanics of Materials

Stresses and Strains 29-1
Thermal, Hoop, and Torsional Stress 30-1
Beams 31-1
Columns 32-1

Topic IX: Electricity and Magnetism

Electrostatics 33-1
Direct-Current Circuits 34-1
Alternating-Current Circuits 35-1
Rotating Machines 36-1

Topic X: Dynamics, Kinematics, and Vibrations

Kinematics 37-1
Kinetics 38-1
Kinetics of Rotational Motion 39-1
Energy and Work 40-1
Vibrations 41-1

Topic XI: Mechanical Design and Analysis

Fasteners 42-1
Machine Design 43-1
Hydraulic and Pneumatic Mechanisms 44-1
Pressure Vessels 45-1
Manufacturability, Quality, and Reliability 46-1

Topic XII: Measurement, Instrumentation, and Controls

Measurement and Instrumentation 47-1
Controls 48-1

Topic XIII: Computational Tools

Computer Software 49-1

Topic XIV: Engineering Economics

Engineering Economics 50-1

Topic XV: Ethics and Professional Practice

Professional Practice 51-1
Ethics 52-1
Licensure 53-1

Preface

The purpose of this book is to prepare you for the National Council of Examiners for Engineering and Surveying (NCEES) fundamentals of engineering (FE) exam.

In 2014, the NCEES adopted revised specifications for the exam. The council also transitioned from a paper-based version of the exam to a computer-based testing (CBT) version. The FE exam now requires you to sit in front of a monitor, respond to questions served up by the CBT system, access an electronic reference document, and perform your scratch calculations on a reusable notepad. You may also use an on-screen calculator with which you will likely be unfamiliar. The experience of taking the FE exam will probably be unlike anything you have ever, or will ever again, experience in your career. Similarly, preparing for the exam will be unlike preparing for any other exam.

The CBT FE exam presented three new challenges to me when I began preparing instructional material for it. (1) The subjects in the testable body of knowledge are oddly limited and do not represent a complete cross section of the traditional engineering fundamentals subjects. (2) The NCEES *FE Reference Handbook* (*NCEES Handbook*) is poorly organized, awkwardly formatted, inconsistent in presentation, and idiomatic in convention. (3) Traditional studying, doing homework while working towards a degree, and working at your own desk as a career engineer are poor preparations for the CBT exam experience.

No existing exam review book overcomes all of these challenges. But, I wanted you to have something that does. So, in order to prepare you for the CBT FE exam, this book was designed and written from the ground up. In many ways, this book is as unconventional as the exam.

This book covers all of the knowledge areas listed in the NCEES Mechanical FE exam specifications. And, with the exceptions listed in "How to Use This Book," for better or worse, this book duplicates the terms and variables of the *NCEES Handbook* equations.

NCEES has selected, what it believes to be, all of the engineering fundamentals important to an early-career, minimally-qualified engineer, and has distilled them into its single reference, the *NCEES Handbook*. Personally, I cannot accept the premise that engineers learn and use so little engineering while getting their degrees and during their first few career years. However, regardless of whether you accept the NCEES subset of engineering fundamentals, one thing is certain: In serving as your sole source of formulas, theory, methods, and data during the exam, the *NCEES Handbook* severely limits the types of questions that can be included in the FE exam.

The obsolete paper-based exam required very little knowledge outside of what was presented in the previous editions of the *NCEES Handbook*. That *NCEES Handbook* supported a plug-and-chug examinee performance within a constrained body of knowledge. Based on the current FE exam specifications and the *NCEES Handbook*, the CBT FE exam is even more limited than the old paper-based exam. The number (breadth) of knowledge areas, the coverage (depth) of knowledge areas, the number of questions, and the duration of the exam are all significantly reduced. If you are only concerned about passing and/or "getting it over with" before graduation, these reductions are all in your favor. Your only deterrents will be the cost of the exam and the inconvenience of finding a time and place to take it.

Accepting that "it is what it is," I designed this book to guide you through the exam's body of knowledge.

I have several admissions to make: (1) This book contains nothing magical or illicit. (2) This book, by itself, is only one part of a complete preparation. (3) This book stops well short of being perfect. What do I mean by those admissions?

First, this book does not contain anything magical. It's called a "practice problems" book, and though it will save you time in assembling hundreds of practice problems for your review, it won't learn the material for you. Merely owning it is not enough. You will have to put in the "practice" time to use it.

Similarly, there is nothing clandestine or unethical about this book. It does not contain any actual exam questions. It was written in a vacuum, based entirely on the NCEES Mechanical FE exam specifications. This book is not based on feedback from actual examinees.

Truthfully, I expect that many exam questions will be similar to the questions I have used because NCEES and I developed content with the same set of constraints. (If anything, NCEES is even more constrained when it comes to fringe, outlier, eccentric, original topics.) There is a finite number of ways that questions about Ohm's law ($V = IR$) and Newton's second law of motion ($F = ma$) can be structured. Any similarity between questions in this book and questions in the exam is easily attributed to the limited number of engineering formulas and concepts, the shallowness of coverage, and the need to keep the entire solution process (reading, researching, calculating, and responding) to less than three minutes for each question.

Let me give an example to put some flesh on the bones. As any competent engineer can attest, in order to calculate the pressure drop in a pipe network, you would normally have to (1) determine fluid density and viscosity based on the temperature, (2) convert the mass flow rate to a volumetric flow rate, (3) determine the pipe diameter from the pipe size designation (e.g., pipe schedule), (4) calculate the internal pipe area, (5) calculate the flow velocity, (6) determine the specific roughness from the conduit material, (7) calculate the relative roughness, (8) calculate the Reynolds number, (9) calculate or determine the friction factor graphically, (10) determine the equivalent length of fittings and other minor losses, (11) calculate the head loss, and finally, (12) convert the head loss to pressure drop. Length, flow quantity, and fluid property conversions typically add even more complexity. (SSU viscosity? Diameter in inches? Flow rate in SCFM?) As reasonable and conventional as that solution process is, a question of such complexity is beyond the upper time limit for an FE question.

To make it possible to be solved in the time allowed, any exam question you see is likely to be more limited. In fact, most or all of the information you need to answer a question will be given to you in its question statement. If only the real world were so kind!

Second, by itself, this book is inadequate. It was never intended to define the entirety of your preparation activity. While it introduces problems covering essentially all of the exam knowledge areas and content in the *NCEES Handbook*, an introduction is only an introduction. To be a thorough review, this book needs augmentation.

By design, this book has four significant inadequacies.

1. This book is "only" 231 pages long, so it cannot contain enough of everything for everyone. The number of practice problems that can fit in it are limited. The number of questions needed by you, personally, to come up to speed in a particular subject may be inadequate. For example, how many questions will you have to review in order to feel comfortable about divergence, curl, differential equations, and linear algebra? (Answer: Probably more than are in all of the books you will ever own!) So, additional exposure is inevitable if you want to be adequately prepared in every subject.

2. This book does not contain the *NCEES Handbook*. This book is limited in helping you become familiar with the idiosyncratic sequencing, formatting, variables, omissions, and presentation of topics in the *NCEES Handbook*. The only way to remedy this is to obtain your own copy of the *NCEES Handbook* (available in printed format from PPI and as a free download from the NCEES website) and use it in conjunction with your review.

3. This book is not a practice examination (mock exam, sample exam, etc.). With the advent of the CBT format, any sample exam in printed format is little more than another collection of practice questions. The actual FE exam is taken sitting in front of a computer using an online reference book, so the only way to practice is to sit in front of a computer while you answer questions. Using an online reference is very different from the work environment experienced by most engineers, and it will take some getting used to.

4. This book does not contain explanatory background information, including figures and tables of data. Though all problems have associated step-by-step solutions, these solutions will not teach you the underlying engineering principles you need to solve the problems. Trying to extrapolate engineering principles from the solutions is like reading the ending of a book and then trying to guess at the "whos, whats, wheres, whens, and hows." In other words, reviewing solutions is only going to get you so far if you don't understand a topic. To truly understand how to solve practice problems in topics you're unfamiliar with, you'll need an actual review manual like the one PPI publishes, the *FE Mechanical Review Manual*. In it, you'll find all the "whos and whats" you were previously missing and these problems' "endings" will make much more sense.

Third, and finally, I reluctantly admit that I have never figured out how to write or publish a completely flawless first (or, even subsequent) edition. The PPI staff comes pretty close to perfection in the areas of design, editing, typography, and illustrating. Subject matter experts help immensely with calculation checking. And, beta testing before you see a book helps smooth out wrinkles. However, I still manage to muck up the content. So, I hope you will "let me have it" when you find my mistakes. PPI has established an easy way for you to report an error, as well as to review changes that resulted from errors that others have submitted. Just go to **ppi2pass.com/errata**. When you submit something, I'll receive it via email. When I answer it, you'll receive a response. We'll both benefit.

Best wishes in your examination experience. Stay in touch!

Michael R. Lindeburg, PE

Acknowledgments

Developing a book specific to the computerized Mechanical FE exam has been a monumental project. It involved the usual (from an author's and publisher's standpoint) activities of updating and repurposing existing content and writing new content. However, the project was made extraordinarily more difficult by two factors: (1) a new book design, and (2) the publication schedule.

Creating a definitive resource to help you prepare for the computerized FE exam was a huge team effort, and PPI's entire Product Development and Implementation (PD&I) staff was heavily involved. Along the way, they had to learn new skills and competencies, solve unseen technical mysteries, and exercise professional judgment in decisions that involved publishing, resources, engineering, and user utility. They worked long hours, week after week, and month after month, often into the late evening, to publish this book for examinees taking the exam.

PPI staff members have had a lot of things to say about this book during its development. In reference to you and other examinees being unaware of what PPI staff did, one of the often-heard statements was, "They will never know."

However, I want you to know, so I'm going to tell you.

Editorial project managers Chelsea Logan, Magnolia Molcan, and Julia White managed the gargantuan operation, with considerable support from Sarah Hubbard, director of PD&I. Production services manager Cathy Schrott kept the process moving smoothly and swiftly, despite technical difficulties that seemed determined to stall the process at every opportunity. Christine Eng, product development manager, arranged for all of the outside subject matter experts that were involved with this book. All of the content was eventually reviewed for consistency, PPI style, and accuracy by Jennifer Lindeburg King, associate editor-in-chief.

Though everyone in PD&I has a specialty, this project pulled everyone from his or her comfort zone. The entire staff worked on "building" the chapters of this book from scratch, piecing together existing content with new content. Everyone learned (with amazing speed) how to grapple with the complexities of XML and MathML while wrestling misbehaving computer code into submission. Tom Bergstrom, technical illustrator, and Kate Hayes, production associate, updated existing illustrations and created new ones. They also paginated and made corrections. Copy editors Tyler Hayes, Scott Marley, Connor Sempek, and Ian A. Walker copy edited, proofread, corrected, and paginated. Copy editors Alexander Ahn, Manuel Carreiro, David Chu, Nicole Evans, EIT, Hilary Flood, Julia Lopez, Ellen Nordman, and Heather Turbeville proofread and corrected. Scott's comments were particularly insightful. Nicole Evans, EIT; Prajesh Gongal, EIT; and Jumphol Somsaad assisted with content selection, problem writing, and calculation checking. Jeanette Baker, EIT; Scott Miller, EIT; Alex Valeyev, EIT; and Akira Zamudio, EIT, remapped existing PPI problems to the new NCEES Mechanical FE exam specifications. Staff engineer Phil Luna, PE, helped ensure the technical accuracy of the content.

Paying customers (such as you) shouldn't have to be test pilots. So, close to the end of the process, when content was starting to coalesce out of the shapelessness of the PPI content management system, several subject matter experts became crash car dummies "for the good of engineering." They pretended to be examinees and worked through all of the content, looking for calculation errors, references that went nowhere, and logic that was incomprehensible. These engineers and their knowledge area contributions are: C. Dale Buckner, PhD, PE, SECB (Statics); John C. Crepeau, PhD, PE (Dynamics, Kinematics, and Vibrations; Electricity and Magnetism; Heat Transfer; Mathematics; Measurements, Instrumentation, and Controls; Mechanical Design and Analysis; and Mechanics of Materials); Joshua T. Frohman, PE (Computational Tools and Transportation Engineering); David Hurwitz, PhD (Computational Tools and Probability and Statistics); Liliana M. Kandic, PE (Fluid Mechanics and Statics); Aparna Phadnis, PE (Engineering Economics); David To, PE (Dynamics, Kinematics, and Vibrations; Electricity and Magnetism; Fluid Mechanics; Heat Transfer; Mathematics; Measurements, Instrumentation, and Controls; Mechanical Design and Analysis; Mechanics of Materials; and Thermodynamics); and L. Adam Williamson, PE (Heat Transfer; Fluid Mechanics; Dynamics, Kinematics, and Vibrations; Material Properties and Processing; and Thermodynamics).

Consistent with the past 36 years, I continue to thank my wife, Elizabeth, for accepting and participating in a writer's life that is full to overflowing. Even though our children have been out on their own for a long time, we seem to have even less time than we had before. As a corollary to Aristotle's "Nature abhors a vacuum," I propose: "Work expands to fill the void."

To my granddaughter, Sydney, who had to share her Grumpus with his writing, I say, "I only worked when you were taking your naps. And besides, you hog the bed!"

Thank you, everyone! I'm really proud of what you've accomplished. Your efforts will be pleasing to examinees and effective in preparing them for the Mechanical FE exam.

                        Michael R. Lindeburg, PE

How to Use This Book

This book is written for one purpose, and one purpose only: to get you ready for the FE exam. Because it is a practice problems book, there are a few, but not many, ways to use it. Here's how this book was designed to be used.

GET THE NCEES *FE REFERENCE* HANDBOOK

Get a copy of the NCEES *FE Reference Handbook* (*NCEES Handbook*). Use it as you solve the problems in this book. The *NCEES Handbook* is the only reference you can use during the exam, so you will want to know the sequence of the sections, what data are included, and the approximate locations of important figures and tables in the *NCEES Handbook*. You should also know the terminology (words and phrases) used in the *NCEES Handbook* to describe equations or subjects, because those are the terms you will have to look up during the exam.

The *NCEES Handbook* is available both in printed and PDF format. The index of the print version may help you locate an equation or other information you are looking for, but few terms are indexed thoroughly. The PDF version includes search functionality that is similar to what you'll have available when taking the computer-based exam. In order to find something using the PDF search function, your search term will have to match the content exactly (including punctuation).

There are a few important differences between the ways the *NCEES Handbook* and this book present content. These differences are intentional for the purpose of maintaining clarity and following PPI's publication policies.

- *pressure:* The *NCEES Handbook* primarily uses P for pressure, an atypical engineering convention. This book always uses p so as to differentiate it from P, which is reserved for power, momentum, and axial loading in related chapters.

- *velocity:* The *NCEES Handbook* uses v and occasionally Greek nu, ν, for velocity. This book always uses v to differentiate it from Greek upsilon, υ, which represents specific volume in some topics (e.g., thermodynamics), and Greek nu, ν, which represents absolute viscosity and Poisson's ratio.

- *specific volume:* The *NCEES Handbook* uses v for specific volume. This book always uses Greek upsilon, υ, a convention that most engineers will be familiar with.

- *units:* The *NCEES Handbook* and the FE exam generally do not emphasize the difference between pounds-mass and pounds-force. "Pounds" ("lb") can mean either force or mass. This book always distinguishes between pounds-force (lbf) and pounds-mass (lbm).

WORK THROUGH EVERY PROBLEM

NCEES has greatly reduced the number of subjects about which you are expected to be knowledgeable and has made nothing optional. Skipping your weakest subjects is no longer a viable preparation strategy. You should study all examination knowledge areas, not just your specialty areas. That means you solve every problem in this book and skip nothing. Do not limit the number of problems you solve in hopes of finding enough problems in your areas of expertise to pass the exam.

The FE exam primarily uses SI units. Therefore, the need to work problems in both the customary U.S. and SI systems is greatly diminished. You will need to learn the SI system if you are not already familiar with it.

BE THOROUGH

Being thorough means really doing the work. Some people think they can read a problem statement, think about it for ten seconds, read the solution, and then say, "Yes, that's what I was thinking of, and that's what I would have done." Sadly, these people find out too late that the human brain doesn't learn very efficiently that way. Under pressure, they find they know and remember very little. For real learning, you'll have to spend some time with a stubby pencil.

There are so many places where you can get messed up solving a problem. Maybe it's in the use of your calculator, like pushing log instead of ln, or forgetting to set the angle to radians instead of degrees, and so on. Maybe it's rusty math. What is $\ln(e^x)$ anyway? How do you factor a polynomial? Maybe it's in finding the data needed or the proper unit conversion. Maybe you're not familiar with the SI system of units. These things take time. And you have to make the mistakes once so that you don't make them again.

If you do decide to get your hands dirty and actually work these problems, you'll have to decide how much reliance you place on this book. It's tempting to turn to a solution when you get slowed down by details or stumped by the subject material. It's tempting to want to maximize the number of problems you solve by spending as little time as possible solving them. However, you need to struggle a little bit more than that to really learn the material.

Studying a new subject is analogous to using a machete to cut a path through a dense jungle. By doing the work, you develop pathways that weren't there before. It's a lot different than just looking at the route on a map. You actually get nowhere by looking at a map. But cut the path once, and you're in business until the jungle overgrowth closes in again. So do the problems—all of them. And, don't look at the solutions until you've sweated a little.

1 Analytic Geometry and Trigonometry

PRACTICE PROBLEMS

1. To find the width of a river, a surveyor sets up a transit at point C on one river bank and sights directly across to point B on the other bank. The surveyor then walks along the bank for a distance of 275 m to point A. The angle CAB is 57° 28′.

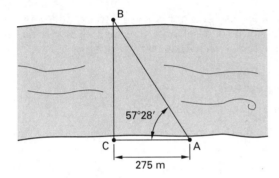

What is the approximate width of the river?

(A) 150 m
(B) 230 m
(C) 330 m
(D) 430 m

2. In the following illustration, angles 2 and 5 are 90°, AD = 15, DC = 20, and AC = 25.

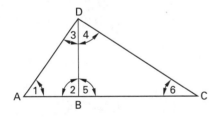

What are the lengths BC and BD, respectively?

(A) 12 and 16
(B) 13 and 17
(C) 16 and 12
(D) 18 and 13

3. What is the length of the line segment with slope 4/3 that extends from the point (6, 4) to the y-axis?

(A) 10
(B) 25
(C) 50
(D) 75

4. Which of the following expressions is equivalent to $\sin 2\theta$?

(A) $2\sin\theta\cos\theta$
(B) $\cos^2\theta - \sin^2\theta$
(C) $\sin\theta\cos\theta$
(D) $\dfrac{1-\cos 2\theta}{2}$

5. Which of the following equations describes a circle with center at $(2,3)$ and passing through the point $(-3,-4)$?

(A) $(x+3)^2 + (y+4)^2 = 85$
(B) $(x+3)^2 + (y+2)^2 = \sqrt{74}$
(C) $(x-3)^2 + (y-2)^2 = 74$
(D) $(x-2)^2 + (y-3)^2 = 74$

6. The equation for a circle is $x^2+4x+y^2+8y=0$. What are the coordinates of the circle's center?

(A) $(-4,-8)$
(B) $(-4,-2)$
(C) $(-2,-4)$
(D) $(2,-4)$

7. Which of the following statements is FALSE for all noncircular ellipses?

(A) The eccentricity, e, is less than one.
(B) The ellipse has two foci.
(C) The sum of the two distances from the two foci to any point on the ellipse is $2a$ (i.e., twice the semimajor distance).
(D) The coefficients A and C preceding the x^2 and y^2 terms in the general form of the equation are equal.

8. What is the area of the shaded portion of the circle shown?

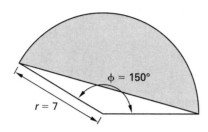

(A) $\dfrac{5\pi}{6} - 1$

(B) $\left(\dfrac{49}{12}\right)(5\pi - 3)$

(C) $\dfrac{50\pi}{3}$

(D) $49\pi - \sqrt{3}$

9. A pipe with a 20 cm inner diameter is filled to a depth equal to one-third of its diameter. What is the approximate area in flow?

(A) 33 cm^2

(B) 60 cm^2

(C) 92 cm^2

(D) 100 cm^2

10. The equation $y = a_1 + a_2 x$ is an algebraic expression for which of the following?

(A) a cosine expansion series

(B) projectile motion

(C) a circle in polar form

(D) a straight line

11. For the right triangle shown, $x = 18$ cm and $y = 13$ cm.

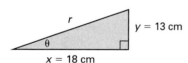

Most nearly, what is $\csc\theta$?

(A) 0.98

(B) 1.2

(C) 1.7

(D) 15

12. A circular sector has a radius of 8 cm and an arc length of 13 cm. Most nearly, what is its area?

(A) 48 cm^2

(B) 50 cm^2

(C) 52 cm^2

(D) 60 cm^2

13. The equation $-3x^2 - 4y^2 = 1$ defines

(A) a circle

(B) an ellipse

(C) a hyperbola

(D) a parabola

14. What is the approximate surface area (including both side and base) of a 4 m high right circular cone with a base 3 m in diameter?

(A) 24 m^2

(B) 27 m^2

(C) 32 m^2

(D) 36 m^2

15. A particle moves in the x-y plane. After t s, the x- and y-coordinates of the particle's location are $x = 8 \sin t$ and $y = 6 \cos t$. Which of the following equations describes the path of the particle?

(A) $36x^2 + 64y^2 = 2304$

(B) $36x^2 - 64y^2 = 2304$

(C) $64x^2 + 36y^2 = 2304$

(D) $64x^2 - 36y^2 = 2304$

SOLUTIONS

1. Use the formula for the tangent of an angle in a right triangle.

$$\tan\theta = BC/AC$$
$$BC = AC\tan\theta = (275 \text{ m})\tan 57°\,28'$$
$$= 431.1 \text{ m} \quad (430 \text{ m})$$

The answer is (D).

2. For right triangle ABD,

$$(BD)^2 + (AB)^2 = (15)^2$$
$$(BD)^2 = (15)^2 - (AB)^2$$

For right triangle DBC,

$$(BD)^2 + (25 - AB)^2 = (20)^2$$
$$(BD)^2 = (20)^2 - (25 - AB)^2$$

Equate the two expressions for $(BD)^2$.

$$(15)^2 - (AB)^2 = (20)^2 - (25)^2 + 50(AB) - (AB)^2$$
$$AB = \frac{(15)^2 - (20)^2 + (25)^2}{50} = 9$$
$$BC = 25 - AB = 25 - 9 = 16$$
$$(BD)^2 = (15)^2 - (9)^2$$
$$BD = 12$$

Alternatively, this problem can be solved using the law of cosines.

The answer is (C).

3. The equation of the line is of the form

$$y = mx + b$$

The slope is $m = 4/3$, and a known point is $(x, y) = (6, 4)$. Find the y-intercept, b.

$$4 = \left(\tfrac{4}{3}\right)(6) + b$$
$$b = 4 - \left(\tfrac{4}{3}\right)(6) = -4$$

The complete equation is

$$y = \tfrac{4}{3}x - 4$$

b is the y-intercept, so the intersection with the y-axis is at point $(0, -4)$. The distance between these two points is

$$d = \sqrt{(y_2 - y_1)^2 + (x_2 - x_1)^2}$$
$$= \sqrt{(4 - (-4))^2 + (6 - 0)^2}$$
$$= 10$$

The answer is (A).

4. The double angle identity is

$$\sin 2\theta = 2\sin\theta\cos\theta$$

The answer is (A).

5. Substitute the known points into the center-radius form of the equation of a circle.

$$r^2 = (x - h)^2 + (y - k)^2$$
$$= (-3 - 2)^2 + (-4 - 3)^2$$
$$= 74$$

The equation of the circle is

$$(x - 2)^2 + (y - 3)^2 = 74$$

($r^2 = 74$. The radius is $\sqrt{74}$.)

The answer is (D).

6. Use the standard form of the equation of a circle to find the circle's center.

$$x^2 + 4x + y^2 + 8y = 0$$
$$x^2 + 4x + 4 + y^2 + 8y + 16 = 4 + 16$$
$$(x + 2)^2 + (y + 4)^2 = 20$$

The center is at $(-2, -4)$.

The answer is (C).

7. The coefficients preceding the squared terms in the general equation are equal only for a straight line or circle, not for a noncircular ellipse.

$$Ax^2 + Bxy + Cy^2 + Dx + Ey + F = 0$$

The answer is (D).

8. The area of the circle is

$$\phi = (150°)\left(\frac{2\pi \text{ rad}}{360°}\right) = \frac{5\pi}{6} \text{ rad}$$
$$A = \frac{r^2(\phi - \sin\phi)}{2}$$
$$= \frac{(7)^2\left(\frac{5\pi}{6} - \sin\frac{5\pi}{6}\right)}{2}$$
$$= \left(\frac{49}{2}\right)\left(\frac{5\pi}{6} - \frac{1}{2}\right)$$
$$= \left(\frac{49}{12}\right)(5\pi - 3)$$

The answer is (B).

9. Find the angle ϕ.

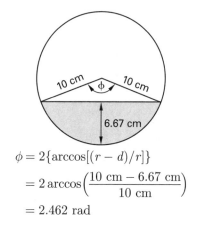

$$\phi = 2\{\arccos[(r-d)/r]\}$$
$$= 2\arccos\left(\frac{10\text{ cm} - 6.67\text{ cm}}{10\text{ cm}}\right)$$
$$= 2.462\text{ rad}$$

Find the area of flow.

$$A = [r^2(\phi - \sin\phi)]/2$$
$$= \frac{(10\text{ cm})^2\Big(2.46\text{ rad} - \sin(2.462\text{ rad})\Big)}{2}$$
$$= 91.67\text{ cm}^2 \quad (92\text{ cm}^2)$$

The answer is (C).

10. $y = mx + b$ is the slope-intercept form of the equation of a straight line. a_1 and a_2 are both constants, so $y = a_1 + a_2 x$ describes a straight line.

The answer is (D).

11. Find the length of the hypotenuse, r.

$$r = \sqrt{x^2 + y^2} = \sqrt{(18\text{ cm})^2 + (13\text{ cm})^2} = 22.2\text{ cm}$$

Find $\csc\theta$.

$$\csc\theta = r/y = \frac{22.2\text{ cm}}{13\text{ cm}} = 1.7$$

The answer is (C).

12. Find the area of the circular sector.

$$A = sr/2 = \frac{(13\text{ cm})(8\text{ cm})}{2} = 52\text{ cm}^2$$

The answer is (C).

13. The general form of the conic section equation is

$$Ax^2 + Bxy + Cy^2 + Dx + Ey + F = 0$$

$A = -3$, $C = -4$, $F = -1$, and $B = D = E = 0$. A and C are different, so the equation does not define a circle. Calculate the discriminant.

$$B^2 - 4AC = (0)^2 - (4)(-3)(-4) = -48$$

This is less than zero, so the equation defines an ellipse.

The answer is (B).

14. Find the total surface area of a right circular cone. The radius is $r = d/2 = 3\text{ m}/2 = 1.5\text{ m}$.

$$A = \text{side area} + \text{base area} = \pi r\left(r + \sqrt{r^2 + h^2}\right)$$
$$= \pi(1.5\text{ m})\left(1.5\text{ m} + \sqrt{(1.5\text{ m})^2 + (4\text{ m})^2}\right)$$
$$= 27.2\text{ m}^2 \quad (27\text{ m}^2)$$

The answer is (B).

15. Rearrange the two coordinate equations.

$$\sin t = \frac{x}{8}$$
$$\cos t = \frac{y}{6}$$

Use the following trigonometric identity.

$$\sin^2\theta + \cos^2\theta = 1$$
$$\left(\frac{x}{8}\right)^2 + \left(\frac{y}{6}\right)^2 = 1$$

To clear the fractions, multiply both sides by $(8)^2 \times (6)^2 = 2304$.

$$36x^2 + 64y^2 = 2304$$

The answer is (A).

2 Algebra and Linear Algebra

PRACTICE PROBLEMS

1. What is the name for a vector that represents the sum of two vectors?

(A) scalar
(B) resultant
(C) tensor
(D) moment

2. The second and sixth terms of a geometric progression are 3/10 and 243/160, respectively. What is the first term of this sequence?

(A) 1/10
(B) 1/5
(C) 3/5
(D) 3/2

3. Using logarithmic identities, what is most nearly the numerical value for the following expression?

$$\log_3 \tfrac{3}{2} + \log_3 12 - \log_3 2$$

(A) 0.95
(B) 1.33
(C) 2.00
(D) 2.20

4. Which of the following statements is true for a power series with the general term $a_i x^i$?

I. An infinite power series converges for $x < 1$.
II. Power series can be added together or subtracted within their interval of convergence.
III. Power series can be integrated within their interval of convergence.

(A) I only
(B) II only
(C) I and III
(D) II and III

5. What is most nearly the length of the resultant of the following vectors?

$$3\mathbf{i} + 4\mathbf{j} - 5\mathbf{k}$$
$$7\mathbf{i} + 2\mathbf{j} + 3\mathbf{k}$$
$$-16\mathbf{i} - 14\mathbf{j} + 2\mathbf{k}$$

(A) 3
(B) 4
(C) 10
(D) 14

6. What is the solution to the following system of simultaneous linear equations?

$$10x + 3y + 10z = 5$$
$$8x - 2y + 9z = 3$$
$$8x + y - 10z = 7$$

(A) $x = 0.326$; $y = -0.192$; $z = 0.586$
(B) $x = 0.148$; $y = 1.203$; $z = 0.099$
(C) $x = 0.625$; $y = 0.186$; $z = -0.181$
(D) $x = 0.282$; $y = -1.337$; $z = -0.131$

7. What is the inverse of matrix $\mathbf{A}$?

$$\mathbf{A} = \begin{bmatrix} 2 & 3 \\ 1 & 1 \end{bmatrix}$$

(A) $\begin{bmatrix} 2 & 3 \\ 1 & 1 \end{bmatrix}$

(B) $\begin{bmatrix} 3 & 2 \\ 1 & 1 \end{bmatrix}$

(C) $\begin{bmatrix} 1 & -3 \\ -1 & 2 \end{bmatrix}$

(D) $\begin{bmatrix} -1 & 3 \\ 1 & -2 \end{bmatrix}$

8. If the determinant of matrix **A** is −40, what is the determinant of matrix **B**?

$$\mathbf{A} = \begin{bmatrix} 4 & 3 & 2 & 1 \\ 0 & 1 & 2 & -1 \\ 2 & 3 & -1 & 1 \\ 1 & 1 & 1 & 2 \end{bmatrix} \quad \mathbf{B} = \begin{bmatrix} 2 & 1.5 & 1 & 0.5 \\ 0 & 1 & 2 & -1 \\ 2 & 3 & -1 & 1 \\ 1 & 1 & 1 & 2 \end{bmatrix}$$

(A) −80

(B) −40

(C) −20

(D) 0.5

9. Given the origin-based vector $\mathbf{A} = \mathbf{i} + 2\mathbf{j} + \mathbf{k}$, what is most nearly the angle between **A** and the x-axis?

(A) 22°

(B) 24°

(C) 66°

(D) 80°

10. Which is a true statement about these two vectors?

$$\mathbf{A} = \mathbf{i} + 2\mathbf{j} + \mathbf{k}$$
$$\mathbf{B} = \mathbf{i} + 3\mathbf{j} - 7\mathbf{k}$$

(A) Both vectors pass through the point $(0, -1, 6)$.

(B) The vectors are parallel.

(C) The vectors are orthogonal.

(D) The angle between the vectors is 17.4°.

11. What is most nearly the acute angle between vectors $\mathbf{A} = (3, 2, 1)$ and $\mathbf{B} = (2, 3, 2)$, both based at the origin?

(A) 25°

(B) 33°

(C) 35°

(D) 59°

12. Force vectors **A**, **B**, and **C** are applied at a single point.

$$\mathbf{A} = \mathbf{i} + 3\mathbf{j} + 4\mathbf{k}$$
$$\mathbf{B} = 2\mathbf{i} + 7\mathbf{j} - \mathbf{k}$$
$$\mathbf{C} = -\mathbf{i} + 4\mathbf{j} + 2\mathbf{k}$$

What is most nearly the magnitude of the resultant force vector, **R**?

(A) 13

(B) 14

(C) 15

(D) 16

13. What is the sum of $12 + 13j$ and $7 - 9j$?

(A) $19 - 22j$

(B) $19 + 4j$

(C) $25 - 22j$

(D) $25 + 4j$

14. What is the product of the complex numbers $3 + 4j$ and $7 - 2j$?

(A) $10 + 2j$

(B) $13 + 22j$

(C) $13 + 34j$

(D) $29 + 22j$

SOLUTIONS

1. By definition, the sum of two vectors is known as the resultant.

The answer is (B).

2. The common ratio is

$$l = ar^{n-1}$$
$$\frac{l_6}{l_2} = \frac{ar^{6-1}}{ar^{2-1}} = r^4$$

$$r = \sqrt[4]{\frac{l_6}{l_2}}$$

$$= \sqrt[4]{\frac{\frac{243}{160}}{\frac{3}{10}}}$$

$$= 3/2$$

The term before $3/10$ is

$$a_1 = \frac{\frac{3}{10}}{\frac{3}{2}} = 1/5$$

The answer is (B).

3. Use the logarithmic identities.

$$\log xy = \log x + \log y$$
$$\log x/y = \log x - \log y$$

$$\log_3 \frac{3}{2} + \log_3 12 - \log_3 2 = \log_3 \frac{\left(\frac{3}{2}\right)(12)}{2}$$
$$= \log_3 9$$

Since $(3)^2 = 9$,

$$\log_3 9 = 2.00$$

The answer is (C).

4. Power series can be added together, subtracted from each other, differentiated, and integrated within their interval of convergence. The interval of convergence is $-1 < x < 1$.

The answer is (D).

5. The resultant is produced by adding the vectors.

$$\begin{array}{r} 3\mathbf{i} + 4\mathbf{j} - 5\mathbf{k} \\ 7\mathbf{i} + 2\mathbf{j} + 3\mathbf{k} \\ -16\mathbf{i} - 14\mathbf{j} + 2\mathbf{k} \\ \hline -6\mathbf{i} - 8\mathbf{j} + 0\mathbf{k} \end{array}$$

The length of the resultant vector is

$$|\mathbf{R}| = \sqrt{(-6)^2 + (-8)^2 + (0)^2}$$
$$= 10$$

The answer is (C).

6. There are several ways of solving this problem.

$$\mathbf{AX} = \mathbf{B}$$

$$\begin{bmatrix} 10 & 3 & 10 \\ 8 & -2 & 9 \\ 8 & 1 & -10 \end{bmatrix} \begin{bmatrix} x \\ y \\ z \end{bmatrix} = \begin{bmatrix} 5 \\ 3 \\ 7 \end{bmatrix}$$

$$\mathbf{AA}^{-1}\mathbf{X} = \mathbf{A}^{-1}\mathbf{B}$$
$$\mathbf{IX} = \mathbf{A}^{-1}\mathbf{B}$$
$$\mathbf{X} = \mathbf{A}^{-1}\mathbf{B}$$

$$\mathbf{X} = \begin{bmatrix} \frac{11}{806} & \frac{20}{403} & \frac{47}{806} \\ \frac{76}{403} & \frac{-90}{403} & \frac{-5}{403} \\ \frac{12}{403} & \frac{7}{403} & \frac{-22}{403} \end{bmatrix} \begin{bmatrix} 5 \\ 3 \\ 7 \end{bmatrix}$$

$$= \begin{bmatrix} (5)\left(\frac{11}{806}\right) + (3)\left(\frac{20}{403}\right) + (7)\left(\frac{47}{806}\right) \\ (5)\left(\frac{76}{403}\right) + (3)\left(\frac{-90}{403}\right) + (7)\left(\frac{-5}{403}\right) \\ (5)\left(\frac{12}{403}\right) + (3)\left(\frac{7}{403}\right) + (7)\left(\frac{-22}{403}\right) \end{bmatrix}$$

$$= \begin{bmatrix} 0.625 \\ 0.186 \\ -0.181 \end{bmatrix}$$

(Direct substitution of the four answer choices into the original equations is probably the fastest way of solving this type of problem.)

The answer is (C).

7. Find the determinant.

$$|\mathbf{A}| = 2 \times 1 - 1 \times 3 = -1$$

The inverse of a 2×2 matrix is

$$\mathbf{A}^{-1} = \frac{\mathrm{adj}(\mathbf{A})}{|\mathbf{A}|} = \frac{\begin{bmatrix} b_2 & -a_2 \\ -b_1 & a_1 \end{bmatrix}}{|\mathbf{A}|}$$

$$= \frac{\begin{bmatrix} 1 & -3 \\ -1 & 2 \end{bmatrix}}{-1}$$

$$= \begin{bmatrix} -1 & 3 \\ 1 & -2 \end{bmatrix}$$

The answer is (D).

8. The first row of matrix **B** is half that of **A**, and the other rows are the same in **A** and **B**, so the determinant of **B** is half the determinant of **A**.

The answer is (C).

9. The magnitude of vector **A** is

$$|\mathbf{A}| = \sqrt{(1)^2 + (2)^2 + (1)^2} = \sqrt{6}$$

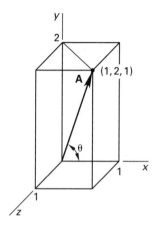

The x-component of the vector is 1, so the direction cosine is

$$\cos\theta_x = \frac{1}{\sqrt{6}}$$

The angle is

$$\theta = \cos^{-1}\left(\frac{1}{\sqrt{6}}\right) = 65.9° \quad (66°)$$

The answer is (C).

10. The magnitudes of the two vectors are

$$|\mathbf{A}| = \sqrt{(1)^2 + (2)^2 + (1)^2} = \sqrt{6}$$
$$|\mathbf{B}| = \sqrt{(1)^2 + (3)^2 + (-7)^2} = \sqrt{59}$$

The angle between them is

$$\phi = \cos^{-1}\left(\frac{a_x b_x + a_y b_y + a_z b_z}{|\mathbf{A}||\mathbf{B}|}\right)$$
$$= \cos^{-1}\left(\frac{(1)(1) + (2)(3) + (1)(-7)}{\sqrt{6}\sqrt{59}}\right)$$
$$= 90°$$

The vectors are orthogonal.

The answer is (C).

11. The angle between the two vectors is

$$\theta = \cos^{-1}\left(\frac{\mathbf{A}\cdot\mathbf{B}}{|\mathbf{A}||\mathbf{B}|}\right)$$
$$= \cos^{-1}\left(\frac{a_x b_x + a_y b_y + a_z b_z}{|\mathbf{A}||\mathbf{B}|}\right)$$
$$= \cos^{-1}\left(\frac{(3)(2) + (2)(3) + (1)(2)}{\sqrt{(3)^2 + (2)^2 + (1)^2}\sqrt{(2)^2 + (3)^2 + (2)^2}}\right)$$
$$= 24.8° \quad (25°)$$

The answer is (A).

12. The magnitude of **R** is

$$|\mathbf{R}| = \sqrt{(1+2-1)^2 + (3+7+4)^2 + (4-1+2)^2}$$
$$= \sqrt{4 + 196 + 25}$$
$$= \sqrt{225}$$
$$= 15$$

The answer is (C).

13. Add the real parts and the imaginary parts of each complex number.

$$(a + jb) + (c + jd) = (a + c) + j(b + d)$$
$$(12 + 13j) + (7 - 9j) = (12 + 7) + j(13 + (-9))$$
$$= 19 + 4j$$

The answer is (B).

14. Use the algebraic distributive law and the equivalency $j^2 = -1$.

$$(a + jb)(c + jd) = (ac - bd) + j(ad + bc)$$
$$(3 + 4j)(7 - 2j) = 21 - 8j^2 + 28j - 6j$$
$$= 21 + 8 + 28j - 6j$$
$$= 29 + 22j$$

The answer is (D).

3 Calculus

PRACTICE PROBLEMS

1. Which of the following is NOT a correct derivative?

(A) $\dfrac{d}{dx}\cos x = -\sin x$

(B) $\dfrac{d}{dx}(1-x)^3 = -3(1-x)^2$

(C) $\dfrac{d}{dx}\dfrac{1}{x} = -\dfrac{1}{x^2}$

(D) $\dfrac{d}{dx}\csc x = -\cot x$

2. What is the derivative, dy/dx, of the expression $x^2 y - e^{2x} = \sin y$?

(A) $\dfrac{2e^{2x}}{x^2 - \cos y}$

(B) $\dfrac{2e^{2x} - 2xy}{x^2 - \cos y}$

(C) $2e^{2x} - 2xy$

(D) $x^2 - \cos y$

3. What is the approximate area bounded by the curves $y = 8 - x^2$ and $y = -2 + x^2$?

(A) 22

(B) 27

(C) 30

(D) 45

4. What are the minimum and maximum values, respectively, of the equation $f(x) = 5x^3 - 2x^2 + 1$ on the interval $[-2, 2]$?

(A) $-47, 33$

(B) $-4, 4$

(C) $0.95, 1$

(D) $0, 0.27$

5. In vector calculus, a gradient is a

I. vector that points in the direction of a general rate of change of a scalar field

II. vector that points in the direction of the maximum rate of change of a scalar field

III. scalar that indicates the magnitude of the rate of change of a vector field in a general direction

IV. scalar that indicates the maximum magnitude of the rate of change of a vector field in any particular direction

(A) I only

(B) II only

(C) I and III

(D) II and IV

6. Which of the illustrations shown represents the vector field, $\mathbf{F}(x, y) = -y\mathbf{i} + x\mathbf{j}$, for nonzero values of x and y?

(A)

(B)

(C)

(D)

7. A peach grower estimates that if he picks his crop now, he will obtain 1000 lugs of peaches, which he can sell at $1.00 per lug. However, he estimates that his crop will increase by an additional 60 lugs of peaches for each week that he delays picking, but the price will drop at a rate of $0.025 per lug per week. In addition, he will experience a spoilage rate of approximately 10 lugs for each week he delays. In order to maximize his revenue, how many weeks should he wait before picking the peaches?

(A) 2 weeks
(B) 5 weeks
(C) 7 weeks
(D) 10 weeks

8. Determine the following indefinite integral.

$$\int \frac{x^3 + x + 4}{x^2} dx$$

(A) $\frac{x}{4} + \ln|x| - \frac{4}{x} + C$

(B) $\frac{-x}{2} + \log x - 8x + C$

(C) $\frac{x^2}{2} + \ln|x| - \frac{2}{x^2} + C$

(D) $\frac{x^2}{2} + \ln|x| - \frac{4}{x} + C$

9. Find dy/dx for the parametric equations given.

$$x = 2t^2 - t$$
$$y = t^3 - 2t + 1$$

(A) $3t^2$
(B) $3t^2/2$
(C) $4t - 1$
(D) $(3t^2 - 2)/(4t - 1)$

10. A two-dimensional function, $f(x, y)$, is defined as

$$f(x, y) = 2x^2 - y^2 + 3x - y$$

What is the direction of the line passing through the point $(1, -2)$ that has the maximum slope?

(A) $4\mathbf{i} + 2\mathbf{j}$
(B) $7\mathbf{i} + 3\mathbf{j}$
(C) $7\mathbf{i} + 4\mathbf{j}$
(D) $9\mathbf{i} - 7\mathbf{j}$

11. Evaluate the following limit.

$$\lim_{x \to 2} \left(\frac{x^2 - 4}{x - 2} \right)$$

(A) 0
(B) 2
(C) 4
(D) ∞

12. If $f(x, y) = x^2 y^3 + xy^4 + \sin x + \cos^2 x + \sin^3 y$, what is $\partial f/\partial x$?

(A) $(2x + y)y^3 + 3\sin^2 y \cos y$
(B) $(4x - 3y^2)xy^2 + 3\sin^2 y \cos y$
(C) $(3x + 4y^2)xy + 3\sin^2 y \cos y$
(D) $(2x + y)y^3 + (1 - 2\sin x)\cos x$

13. What is dy/dx if $y = (2x)^x$?

(A) $(2x)^x(2 + \ln 2x)$
(B) $2x(1 + \ln 2x)^x$
(C) $(2x)^x(\ln 2x^2)$
(D) $(2x)^x(1 + \ln 2x)$

SOLUTIONS

1. Determine each of the derivatives.

$$\frac{d}{dx}\cos x = -\sin x \quad [\text{OK}]$$

$$\frac{d}{dx}(1-x)^3 = (3)(1-x)^2(-1) = (-3)(1-x)^2 \quad [\text{OK}]$$

$$\frac{d}{dx}\frac{1}{x} = \frac{d}{dx}x^{-1} = (-1)(x^{-2}) = \frac{-1}{x^2} \quad [\text{OK}]$$

$$\frac{d}{dx}\csc x = -\cot x \quad [\text{incorrect}]$$

The answer is (D).

2. Since neither x nor y can be extracted from the equality, rearrange to obtain a homogeneous expression in x and y.

$$x^2 y - e^{2x} = \sin y$$
$$f(x,y) = x^2 y - e^{2x} - \sin y = 0$$

Take the partial derivatives with respect to x and y.

$$\frac{\partial f(x,y)}{\partial x} = 2xy - 2e^{2x}$$

$$\frac{\partial f(x,y)}{\partial y} = x^2 - \cos y$$

Use implicit differentiation.

$$\frac{\partial y}{\partial x} = \frac{-\frac{\partial f(x,y)}{\partial x}}{\frac{\partial f(x,y)}{\partial y}} = \frac{2e^{2x} - 2xy}{x^2 - \cos y}$$

The answer is (B).

3. Find the intersection points by setting the two functions equal.

$$-2 + x^2 = 8 - x^2$$
$$2x^2 = 10$$
$$x = \pm\sqrt{5}$$

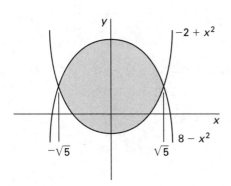

The integral of $f_1(x) - f_2(x)$ represents the area between the two curves between the limits of integration.

$$A = \int_{x_1}^{x_2} (f_1(x) - f_2(x))\,dx$$
$$= \int_{-\sqrt{5}}^{\sqrt{5}} ((8 - x^2) - (-2 + x^2))\,dx$$
$$= \int_{-\sqrt{5}}^{\sqrt{5}} (10 - 2x^2)\,dx$$
$$= \left(10x - \tfrac{2}{3}x^3\right)\Big|_{-\sqrt{5}}^{\sqrt{5}}$$
$$= 29.8 \quad (30)$$

The answer is (C).

4. The critical points are located where the first derivative is zero.

$$f(x) = 5x^3 - 2x^2 + 1$$
$$f'(x) = 15x^2 - 4x$$
$$15x^2 - 4x = 0$$
$$x(15x - 4) = 0$$
$$x = 0 \quad \text{or} \quad x = 4/15$$

Test for a maximum, minimum, or inflection point.

$$f''(x) = 30x - 4$$
$$f''(0) = (30)(0) - 4$$
$$= -4$$
$$f''(a) < 0 \quad [\text{maximum}]$$
$$f''\left(\frac{4}{15}\right) = (30)\left(\frac{4}{15}\right) - 4$$
$$= 4$$
$$f''(a) > 0 \quad [\text{minimum}]$$

These could be a local maximum and minimum. Check the endpoints of the interval and compare with the function values at the critical points.

$$f(-2) = (5)(-2)^3 - (2)(-2)^2 + 1 = -47$$
$$f(2) = (5)(2)^3 - (2)(2)^2 + 1 = 33$$
$$f(0) = (5)(0)^3 - (2)(0)^2 + 1 = 1$$
$$f\left(\frac{4}{15}\right) = (5)\left(\frac{4}{15}\right)^3 - (2)\left(\frac{4}{15}\right)^2 + 1$$
$$= 0.95$$

The minimum and maximum values of the equation, -47 and 33, respectively, are at the endpoints.

The answer is (A).

5. A gradient (gradient vector) at some point P is described by use of the gradient (del, grad, nabla, etc.) function, $\nabla f_P \cdot \mathbf{a}$, where $\mathbf{a}$ is a unit vector. In three-dimensional rectangular coordinates, the gradient is equivalent to the partial derivative vector $\nabla f \cdot \mathbf{a} = \frac{\partial f}{\partial x}\mathbf{i} + \frac{\partial f}{\partial y}\mathbf{j} + \frac{\partial f}{\partial z}\mathbf{k}$. This is a vector that points in the direction of the maximum rate of change (i.e., maximum slope).

The answer is (B).

6. From $-y\mathbf{i}$, it can be concluded that for

(a) positive values of y, the vector field points to the left, and

(b) negative values of y, the vector field points to the right.

From $+x\mathbf{j}$, it can be concluded that for

(a) positive values of x, the vector field points upward, and

(b) negative values of x, the vector field points downward.

The answer is (C).

7. Let x represent the number of weeks.

The equation describing the price as a function of time is

$$\frac{\text{price}}{\text{lug}} = \$1 - \$0.025x$$

The equation describing the yield is

$$\begin{aligned}\text{lugs sold} &= 1000 + (60 - 10)x \\ &= 1000 + 50x\end{aligned}$$

The revenue function is

$$\begin{aligned}R &= \left(\frac{\text{price}}{\text{lug}}\right)(\text{lugs sold}) \\ &= (1 - 0.025x)(1000 + 50x) \\ &= 1000 + 50x - 25x - 1.25x^2 \\ &= 1000 + 25x - 1.25x^2\end{aligned}$$

To maximize the revenue function, set its derivative equal to zero.

$$\frac{dR}{dx} = 25 - 2.5x = 0$$
$$x = 10 \text{ weeks}$$

The answer is (D).

8. Separate the fraction into parts and integrate each one.

$$\begin{aligned}\int \frac{x^3 + x + 4}{x^2}\,dx &= \int \frac{x^3}{x^2}\,dx + \int \frac{x}{x^2}\,dx + \int \frac{4}{x^2}\,dx \\ &= \int x\,dx + \int \frac{1}{x}\,dx + 4\int \frac{1}{x^2}\,dx \\ &= \frac{x^2}{2} + \ln|x| + 4\left(\frac{x^{-1}}{-1}\right) + C \\ &= \frac{x^2}{2} + \ln|x| - \frac{4}{x} + C\end{aligned}$$

The answer is (D).

9. Calculate the derivatives of x and y with respect to t.

$$\frac{dy}{dt} = 3t^2 - 2$$
$$\frac{dx}{dt} = 4t - 1$$

The derivative of y with respect to x is

$$\begin{aligned}\frac{dy}{dx} &= \frac{\frac{dy}{dt}}{\frac{dx}{dt}} \\ &= \frac{3t^2 - 2}{4t - 1}\end{aligned}$$

The answer is (D).

10. The direction of the line passing through $(1, -2)$ with maximum slope is found by inserting $x = 1$ and $y = -2$ into the gradient vector function.

The gradient of the function is

$$\begin{aligned}\nabla f(x, y, z) &= \frac{\partial f(x, y, z)}{\partial x}\mathbf{i} + \frac{\partial f(x, y, z)}{\partial y}\mathbf{j} + \frac{\partial f(x, y, z)}{\partial z}\mathbf{k} \\ &= \frac{\partial(2x^2 - y^2 + 3x - y)}{\partial x}\mathbf{i} \\ &\quad + \frac{\partial(2x^2 - y^2 + 3x - y)}{\partial y}\mathbf{j} \\ &= (4x + 3)\mathbf{i} - (2y + 1)\mathbf{j}\end{aligned}$$

At $(1, -2)$,

$$\begin{aligned}\nabla f(1, -2) &= \big((4)(1) + 3\big)\mathbf{i} - \big((2)(-2) + 1\big)\mathbf{j} \\ &= 7\mathbf{i} + 3\mathbf{j}\end{aligned}$$

The answer is (B).

11. The expression approaches 0/0 at the limit.

$$\frac{(2)^2 - 4}{2 - 2} = \frac{0}{0}$$

Use L'Hôpital's rule.

$$\lim_{x \to 2} \left(\frac{x^2 - 4}{x - 2}\right) = \lim_{x \to 2} \left(\frac{\frac{d}{dx}(x^2 - 4)}{\frac{d}{dx}(x - 2)}\right) = \lim_{x \to 2} \left(\frac{2x}{1}\right)$$

$$= \frac{(2)(2)}{1}$$

$$= 4$$

This could also be solved by factoring the numerator.

The answer is (C).

12. The partial derivative with respect to x is found by treating all other variables as constants. Therefore, all terms that do not contain x have zero derivatives.

$$\frac{\partial f}{\partial x} = 2xy^3 + y^4 + \cos x + 2\cos x(-\sin x)$$

$$= (2x + y)y^3 + (1 - 2\sin x)\cos x$$

The answer is (D).

13. From the table of derivatives,

$$\mathbf{D}\big(f(x)\big)^{g(x)} = g(x)\big(f(x)\big)^{g(x)-1}\mathbf{D}f(x)$$
$$+ \ln(f(x))\big(f(x)\big)^{g(x)}\mathbf{D}g(x)$$

$$f(x) = 2x$$
$$g(x) = x$$

$$\frac{d(2x)^x}{dx} = x(2x)^{x-1}(2) + (\ln 2x)(2x)^x(1)$$

$$= (2x)^x + (2x)^x \ln 2x$$

$$= (2x)^x (1 + \ln 2x)$$

The answer is (D).

4 Differential Equations and Transforms

PRACTICE PROBLEMS

1. What is the solution to the following differential equation?

$$y' + 5y = 0$$

(A) $y = 5x + C$
(B) $y = Ce^{-5x}$
(C) $y = Ce^{5x}$
(D) either (A) or (B)

2. What is the solution to the following linear difference equation?

$$(k+1)(y(k+1)) - ky(k) = 1$$

(A) $y(k) = 12 - \dfrac{1}{k}$
(B) $y(k) = 1 - \dfrac{12}{k}$
(C) $y(k) = 12 + 3k$
(D) $y(k) = 3 + \dfrac{1}{k}$

3. What is the general solution to the following differential equation?

$$2\left(\dfrac{d^2y}{dx^2}\right) - 4\left(\dfrac{dy}{dx}\right) + 4y = 0$$

(A) $y = C_1 \cos x + C_2 \sin x$
(B) $y = C_1 e^x + C_2 e^{-x}$
(C) $y = e^{-x}(C_1 \cos x - C_2 \sin x)$
(D) $y = e^x(C_1 \cos x + C_2 \sin x)$

4. What is the general solution to the following differential equation?

$$\dfrac{d^2y}{dx^2} + 2\dfrac{dy}{dx} + 2y = 0$$

(A) $y = C_1 \sin x - C_2 \cos x$
(B) $y = C_1 \cos x - C_2 \sin x$
(C) $y = C_1 \cos x + C_2 \sin x$
(D) $y = e^{-x}(C_1 \cos x + C_2 \sin x)$

5. What is the complementary solution to the following differential equation?

$$y'' - 4y' + \tfrac{25}{4}y = 10 \cos 8x$$

(A) $y = 2C_1 x + C_2 x - C_3 x$
(B) $y = C_1 e^{2x} + C_2 e^{1.5x}$
(C) $y = C_1 e^{2x} \cos 1.5x + C_2 e^{2x} \sin 1.5x$
(D) $y = C_1 e^x \tan x + C_2 e^x \cot x$

6. What is the general solution to the following differential equation?

$$y'' + y' + y = 0$$

(A) $y = e^{-\frac{1}{2}x}\left(C_1 \cos \tfrac{\sqrt{3}}{2}x + C_2 \sin \tfrac{\sqrt{3}}{2}x\right)$
(B) $y = e^{-\frac{1}{2}x}(C_1 \cos \tfrac{3}{2}x + C_2 \sin \tfrac{3}{2}x)$
(C) $y = e^{-2x}\left(C_1 \cos \tfrac{\sqrt{3}}{2}x + C_2 \sin \tfrac{\sqrt{3}}{2}x\right)$
(D) $y = e^{-2x}(C_1 \cos \tfrac{3}{2}x + C_2 \sin \tfrac{3}{2}x)$

7. What is the solution to the following differential equation if $x = 1$ at $t = 0$, and $dx/dt = 0$ at $t = 0$?

$$\tfrac{1}{2}\dfrac{d^2x}{dt^2} + 4\dfrac{dx}{dt} + 8x = 5$$

(A) $x = e^{-4t} + 4te^{-4t}$
(B) $x = \tfrac{3}{8}e^{-2t}(\cos 2t + \sin 2t) + \tfrac{5}{8}$
(C) $x = e^{-4t} + 4te^{-4t} + \tfrac{5}{8}$
(D) $x = \tfrac{3}{8}e^{-4t} + \tfrac{3}{2}te^{-4t} + \tfrac{5}{8}$

8. In the following differential equation with the initial condition $x(0) = 12$, what is the value of $x(2)$?

$$\dfrac{dx}{dt} + 4x = 0$$

(A) 3.4×10^{-3}
(B) 4.0×10^{-3}
(C) 5.1×10^{-3}
(D) 6.2×10^{-3}

9. What are the three general Fourier coefficients for the sawtooth wave shown?

(A) $a_0 = 0$, $a_n = 0$, $b_n = \dfrac{-1}{\pi n}$

(B) $a_0 = \dfrac{1}{2}$, $a_n = 0$, $b_n = \dfrac{-1}{\pi n}$

(C) $a_0 = 1$, $a_n = 1$, $b_n = \dfrac{1}{\pi n}$

(D) $a_0 = \dfrac{1}{2}$, $a_n = \dfrac{1}{2}$, $b_n = \dfrac{1}{\pi n}$

10. The values of an unknown function follow a Fibonacci number sequence. It is known that $f(1) = 4$ and $f(2) = 1.3$. What is $f(4)$?

(A) −4.1

(B) 0.33

(C) 2.7

(D) 6.6

SOLUTIONS

1. This is a first-order linear equation with characteristic equation $r + 5 = 0$. The form of the solution is

$$y = Ce^{-5x}$$

In the preceding equation, the constant, C, could be determined from additional information.

The answer is (B).

2. Since nothing is known about the general form of $y(k)$, the only way to solve this problem is by trial and error, substituting each answer option into the equation in turn. Option B is

$$y(k) = 1 - \dfrac{12}{k}$$

Substitute this into the difference equation.

$$(k+1)\big(y(k+1)\big) - k\big(y(k)\big) = 1$$
$$(k+1)\left(1 - \dfrac{12}{k+1}\right) - k\left(1 - \dfrac{12}{k}\right) = 1$$
$$(k+1)\left(\dfrac{k+1-12}{k+1}\right) - k\left(\dfrac{k-12}{k}\right) = 1$$
$$k + 1 - 12 - k + 12 = 1$$
$$1 = 1$$

$y(k) = 1 - 12/k$ solves the difference equation.

The answer is (B).

3. This is a second-order, homogeneous, linear differential equation. Start by putting it in general form.

$$y'' + 2ay' + by = 0$$
$$2y'' - 4y' + 4y = 0$$
$$y'' - 2y' + 2y = 0$$
$$a = -2$$
$$b = 2$$

Since $a^2 < 4b$, the form of the equation is

$$y = e^{\alpha x}(C_1 \cos \beta x + C_2 \sin \beta_x)$$
$$\alpha = \dfrac{-a}{2} = \dfrac{-2}{2} = 1$$
$$\beta = \dfrac{\sqrt{4b - a^2}}{2}$$
$$= \dfrac{\sqrt{(4)(2) - (-2)^2}}{2}$$
$$= 1$$
$$y = e^x(C_1 \cos x + C_2 \sin x)$$

The answer is (D).

4. The characteristic equation is

$$r^2 + 2r + 2 = 0$$
$$a = 2$$
$$b = 2$$

The roots are

$$r_{1,2} = \frac{-a \pm \sqrt{a^2 - 4b}}{2}$$
$$= \frac{-2 \pm \sqrt{(2)^2 - (4)(2)}}{2}$$
$$= (-1 + i), (-1 - i)$$

Since $a^2 < 4b$, the solution is

$$y = e^{\alpha x}(C_1 \cos \beta x + C_2 \sin \beta x)$$
$$\alpha = \frac{-a}{2} = \frac{-2}{2} = -1$$
$$\beta = \frac{\sqrt{4b - a^2}}{2} = \frac{\sqrt{(4)(2) - (2)^2}}{2}$$
$$= 1$$
$$y = e^{-x}(C_1 \cos x + C_2 \sin x)$$

The answer is (D).

5. The complementary solution to a nonhomogeneous differential equation is the solution of the homogeneous differential equation.

The characteristic equation is

$$r^2 + ar + b = 0$$
$$r^2 - 4r + \frac{25}{4} = 0$$

So, $a = -4$, and $b = 25/4$.

The roots are

$$r_{1,2} = \frac{-a \pm \sqrt{a^2 - 4b}}{2}$$
$$= \frac{-(-4) \pm \sqrt{(-4)^2 - (4)\left(\frac{25}{4}\right)}}{2}$$
$$= 2 \pm 1.5i$$

Since the roots are imaginary, the homogeneous solution has the form of

$$y = e^{\alpha x}(C_1 \cos \beta x + C_2 \sin \beta x)$$
$$\alpha = 2$$
$$\beta = \pm 1.5$$

The complementary solution is

$$y = e^{2x}(C_1 \cos 1.5x + C_2 \sin 1.5x)$$
$$= C_1 e^{2x} \cos 1.5x + C_2 e^{2x} \sin 1.5x$$

The answer is (C).

6. This is a second-order, homogeneous, linear differential equation with $a = b = 1$. This differential equation can be solved by the method of undetermined coefficients with a solution in the form $y = Ce^{rx}$. The substitution of the solution gives

$$(r^2 + ar + b)Ce^{rx} = 0$$

Because Ce^{rx} can never be zero, the characteristic equation is

$$r^2 + ar + b = 0$$

Because $a^2 = 1 < 4b = 4$, the general solution is in the form

$$y = e^{\alpha x}(C_1 \cos \beta x + C_2 \sin \beta x)$$

Then,

$$\alpha = -a/2 = -1/2$$
$$\beta = \sqrt{\frac{4b - a^2}{2}} = \sqrt{\frac{(4)(1) - (1)^2}{2}} = \frac{\sqrt{3}}{2}$$

Therefore, the general solution is

$$y = e^{-\frac{1}{2}x}\left(C_1 \cos \tfrac{\sqrt{3}}{2}x + C_2 \sin \tfrac{\sqrt{3}}{2}x\right)$$

The answer is (A).

7. Multiplying the equation by 2 gives

$$x'' + 8x' + 16x = 10$$

The characteristic equation is

$$r^2 + 8r + 16 = 0$$

The roots of the characteristic equation are

$$r_1 = r_2 = -4$$

The homogeneous (natural) response is

$$x_{\text{natural}} = Ae^{-4t} + Bte^{-4t}$$

By inspection, $x = 5/8$ is a particular solution that solves the nonhomogeneous equation, so the total response is

$$x = Ae^{-4t} + Bte^{-4t} + \frac{5}{8}$$

Since $x = 1$ at $t = 0$,

$$1 = Ae^0 + \frac{5}{8}$$

$$A = \frac{3}{8}$$

Differentiating x,

$$x' = \frac{3}{8}(-4)e^{-4t} + B(-4te^{-4t} + e^{-4t}) + 0$$

Since $x' = 0$ at $t = 0$,

$$0 = -\frac{3}{2} + B(0 + 1)$$

$$B = \frac{3}{2}$$

$$x = \frac{3}{8}e^{-4t} + \frac{3}{2}te^{-4t} + \frac{5}{8}$$

The answer is (D).

8. This is a first-order, linear, homogeneous differential equation with characteristic equation $r + 4 = 0$.

$$x' + 4x = 0$$

$$x = x_0 e^{-4t}$$

$$x(0) = x_0 e^{(-4)(0)}$$

$$= 12$$

$$x_0 = 12$$

$$x = 12 e^{-4t}$$

$$x(2) = 12 e^{(-4)(2)}$$

$$= 12 e^{-8}$$

$$= 4.03 \times 10^{-3} \quad (4.0 \times 10^{-3})$$

The answer is (B).

9. By inspection, $f(t) = t$, with the period $T = 1$. The angular frequency is

$$\omega_0 = \frac{2\pi}{T} = \frac{2\pi}{1} = 2\pi$$

The average is

$$a_0 = (1/T)\int_0^T f(t)\,dt = (1/T)\int_0^T t\,dt = \tfrac{1}{2}t^2\Big|_0^1 = \frac{1}{2} - 0$$

$$= \frac{1}{2}$$

The general a term is

$$a_n = (2/T)\int_0^T f(t)\cos(n\omega_0 t)\,dt$$

$$= 2\int_0^1 t\cos(2\pi nt)\,dt$$

$$= 0$$

The general b term is

$$b_n = (2/T)\int_0^T f(t)\sin(n\omega_0 t)\,dt$$

$$= 2\int_0^1 t\sin(2\pi nt)\,dt$$

$$= \frac{-1}{\pi n}$$

The answer is (B).

10. The value of a number in a Fibonacci sequence is the sum of the previous two numbers in the sequence.

Use the second-order difference equation.

5 Numerical Methods

PRACTICE PROBLEMS

1. The given section of pseudocode approximates the integral of some function, $F(x)$, over the interval from a to b. Assume that a separate subroutine has already been written to calculate the value of $F(x)$.

```
1   INPUT a, b, n
2   d = (b − a)/n
3   S = F(a)
4   FOR k FROM 1 TO (n − 1)
5   S = S + F(a + k*d)
6   NEXT k
7   S = S*d
```

Which numerical method does this code represent?

(A) Euler's method

(B) Newton's method

(C) Simpson's rule

(D) trapezoidal rule

2. Using Newton's method with an initial estimate of $x_0 = 2$ through two iterations, what is most nearly the root of the following equation?

$$f(x) = x^3 - 4x - 5$$

(A) 2.47

(B) 2.60

(C) 2.63

(D) 16.7

3. Use Euler's approximation to determine $x(1.5)$, the value of a function at $t = 1.5$, given increments of t of 0.25, $x(1) = 1$, and $dx/dt = 2x$.

(A) 0

(B) 0.75

(C) 1.5

(D) 2.3

4. The given section of pseudocode approximates the integral of some function, $F(x)$, over the interval from a to b. Assume that a separate subroutine has already been written to calculate the value of $F(x)$.

```
1   INPUT a, b, n
2   d = (b − a)/n
3   S = F(a)/2 + F(b)/2
4   FOR k FROM 1 TO (n − 1)
5   S = S + F(a + k*d)
6   NEXT k
7   S = S*d
```

Using the pseudocode's algorithm with $n = 4$, what would be the approximation of $\int_0^\pi \sin x \, dx$?

(A) 1.4

(B) 1.6

(C) 1.9

(D) 2.0

5. Using two iterations of Newton's method and with an initial estimate of $x_0 = 2$, what is most nearly the root of the following equation?

$$f(x) = x^3 - 2x - 7$$

(A) 2.00

(B) 2.26

(C) 2.30

(D) 2.34

SOLUTIONS

1. The pseudocode compiles a sum, S, which is used to approximate the integral. Starting with line 3,

$$S = F(a)$$

The FOR...NEXT loop (lines 4, 5, and 6) is effectively a summation.

$$S = S + \sum_{k=1}^{n-1} F(a+kd) = F(a) + \sum_{k=1}^{n-1} F(a+kd)$$

$$= \sum_{k=0}^{n-1} F(a+kd)$$

Finally, line 7 multiplies the previous sum by d.

$$S = d\left(\sum_{k=0}^{n-1} F(a+kd)\right)$$

This is an implementation of Euler's method.

The answer is (A).

2. Newton's method for finding a root is

$$a_{j+1} = a_j - \frac{f(x_n)}{f'(x_n)}$$

The function and its first derivative are

$$f(x) = x^3 - 4x - 5$$
$$f'(x) = 3x^2 - 4$$

The first iteration, with $n=0$, results in

$$x_0 = a_0 = 2$$
$$f(x_0) = f(2) = (2)^3 - (4)(2) - 5 = 5$$
$$f'(x_0) = f'(2) = (3)(2)^2 - 4 = 8$$
$$a_1 = a_0 - \frac{f(x_0)}{f'(x_0)} = 2 - \frac{-5}{8}$$
$$= 2.625$$

The second iteration, with $n=1$, results in

$$x_1 = a_1 = 2.625$$
$$f(x_1) = f(2.625) = (2.625)^3 - (4)(2.625) - 5 = 2.588$$
$$f'(x_1) = f'(2.625) = (3)(2.625)^2 - 4 = 16.672$$
$$a_2 = a_1 - \frac{f(x_1)}{f'(x_1)} = 2.625 - \frac{2.588}{16.672}$$
$$= 2.470 \quad (2.47)$$

The answer is (A).

3. Euler's approximation is

$$x_{k+1} = x_k + \Delta t \frac{dx_k}{dt}$$

$$x(1+0.25) = x(1) + \Delta t \Big(2x(1)\Big)$$

$$x(1.25) = 1 + (0.25)(2)(1) = 1.5$$

$$x(1.25+0.25) = x(1.25) + \Delta t \Big(2x(1.25)\Big)$$

$$x(1.5) = 1.5 + (0.25)(2)(1.5)$$

$$= 2.25 \quad (2.3)$$

The answer is (D).

4. Use the algorithm provided to approximate the integral.

$$d = \frac{b-a}{2} = \frac{\pi - 0}{4} = \frac{\pi}{4}$$

The integral is equal to approximately

$$S = d\left(\frac{F(a)}{s} + \frac{F(b)}{2} + F(a+d) + F(a+2d) + F(a+3d)\right)$$

$$= \left(\frac{\pi}{4}\right)\left(\frac{F(0)}{2} + \frac{F(\pi)}{2} + F\left(\frac{\pi}{4}\right) + F\left(\frac{\pi}{2}\right) + F\left(\frac{3\pi}{4}\right)\right)$$

$$= \left(\frac{\pi}{4}\right)\left(0 + 0 + \frac{\sqrt{2}}{2} + 1 + \frac{\sqrt{2}}{2}\right)$$

$$= 1.896 \quad (1.9)$$

The answer is (C).

5. The function and its first derivative are

$$f(x) = x^3 - 2x - 7$$
$$\frac{df(x)}{dx} = 3x^2 - 2$$

For the first iteration, $n=0$.

$$a_0 = 2$$
$$f(x_0) = f(2) = (2)^3 - (2)(2) - 7 = -3$$
$$\frac{df(x_0)}{dx} = \frac{df(2)}{dx} = (3)(2)^2 - 2 = 10$$
$$a_1 = a_0 - \frac{f(x_0)}{\frac{df(x_0)}{dx}} = 2 - \frac{-3}{10} = 2.3$$

For the second iteration, $n=1$.

$$a_1 = 2.3$$
$$f(x_1) = (2.3)^3 - (2)(2.3) - 7$$
$$= 0.567$$
$$\frac{df(x_1)}{dx} = (3)(2.3)^2 - 2$$
$$= 13.87$$
$$a_2 = a_1 - \frac{f(x_1)}{\frac{df(x_1)}{dx}}$$
$$= 2.3 - \frac{0.567}{13.87}$$
$$= 2.259 \quad (2.26)$$

The answer is (B).

6 Probability and Statistics

PRACTICE PROBLEMS

1. What is the approximate probability that no people in a group of seven have the same birthday?

(A) 0.056
(B) 0.43
(C) 0.92
(D) 0.94

2. A study gives the following results for a total sample size of 12.

$$3, 4, 4, 5, 8, 8, 8, 10, 11, 15, 18, 20$$

What is most nearly the mean?

(A) 8.9
(B) 9.5
(C) 11
(D) 12

3. A study gives the following results for a total sample size of 8.

$$2, 3, 5, 8, 8, 10, 10, 12$$

The mean of the sample is 7.25. What is most nearly the standard deviation?

(A) 2.5
(B) 2.9
(C) 3.3
(D) 3.7

4. A study gives the following results for a total sample size of 6.

$$10, 12, 13, 14, 14, 15$$

The mean of the sample is 13. What is most nearly the sample standard deviation?

(A) 0.85
(B) 0.90
(C) 1.6
(D) 1.8

5. A study has a sample size of 5, a standard deviation of 10.4, and a sample standard deviation of 11.6. What is most nearly the variance?

(A) 46
(B) 52
(C) 110
(D) 130

6. A study has a sample size of 9, a standard deviation of 4.0, and a sample standard deviation of 4.2. What is most nearly the sample variance?

(A) 16
(B) 18
(C) 34
(D) 36

7. A bag contains 100 balls numbered 1 to 100. One ball is drawn from the bag. What is the probability that the number on the ball selected will be odd or greater than 80?

(A) 0.1
(B) 0.5
(C) 0.6
(D) 0.7

8. Measurements of the water content of soil from a borrow site are normally distributed with a mean of 14.2% and a standard deviation of 2.3%. What is the probability that a sample taken from the site will have a water content above 16% or below 12%?

(A) 0.13
(B) 0.25
(C) 0.37
(D) 0.42

9. What is the probability that either exactly two heads or exactly three heads will be thrown if six fair coins are tossed at once?

(A) 0.35
(B) 0.55
(C) 0.59
(D) 0.63

10. Which of the following properties of probability is NOT valid?

(A) The probability of an event is always positive and within the range of zero and one.

(B) The probability of an event which cannot occur in the population being examined is zero.

(C) If events A and B are mutually exclusive, then the probability of either event occurring in the same population is zero.

(D) The probability of either of two events, A and B, occurring is $P(A+B) = P(A) + P(B) - P(A,B)$.

11. One fair die is used in a dice game. A player wins $10 if he rolls either a 1 or a 6. He loses $5 if he rolls any other number. What is the expected winning for one roll of the die?

(A) $0.00
(B) $3.30
(C) $5.00
(D) $6.70

12. A simulation model for a transportation system is run for 30 replications, and the mean percentage utilization of the transporter used by the system is recorded for each replication. Those 30 data points are then used to form a confidence interval on mean transporter utilization for the system. At a 95% confidence level, the confidence interval is found to be 37.2% ± 3.4%.

Given this information, which of the following facts can be definitively stated about the system?

(A) At 95% confidence, the sample mean of transporter utilization lies in the range 37.2% ± 3.4%.

(B) At 95% confidence, the population mean of transporter utilization lies in the range 37.2% ± 3.4%.

(C) At 95% confidence, the population mean of transporter utilization lies outside of the range of 37.2% ± 3.4%.

(D) At 5% confidence, the population mean of transporter utilization lies inside of the range of 37.2% ± 3.4%.

13. What is the approximate probability of exactly two people in a group of seven having a birthday on April 15?

(A) 1.2×10^{-18}
(B) 2.4×10^{-17}
(C) 7.4×10^{-6}
(D) 1.6×10^{-4}

14. What are the arithmetic mean and sample standard deviation of the following numbers?

$$71.3, 74.0, 74.25, 78.54, 80.6$$

(A) 74.3, 2.7
(B) 74.3, 3.7
(C) 75.0, 2.7
(D) 75.7, 3.8

15. Four fair coins are tossed at once. What is the probability of obtaining three heads and one tail?

(A) 1/4 (0.25)
(B) 3/8 (0.375)
(C) 1/2 (0.50)
(D) 3/4 (0.75)

16. Set A and set B are subsets of set Q. The values within each set are shown.

$$A = (4, 7, 9)$$
$$B = (4, 5, 9, 10)$$
$$Q = (4, 5, 6, 7, 8, 9, 10)$$

What is the union of the complement of set A and set B, $\overline{A} \cup B$?

(A) (4, 5, 6, 7, 8, 9, 10)
(B) (4, 5, 7, 9, 10)
(C) (4, 5, 6, 8, 9, 10)
(D) (5, 10)

17. Set A consists of elements (1, 3, 6), and set B consists of elements (1, 2, 6, 7). Both sets come from the universe of (1, 2, 3, 4, 5, 6, 7, 8). What is the intersection, $\overline{A} \cap B$?

(A) (2, 7)
(B) (2, 3, 7)
(C) (2, 4, 5, 7, 8)
(D) (4, 5, 8)

SOLUTIONS

1. This is the classic "birthday problem." The problem is to find the probability that all seven people have distinctly different birthdays. The solution can be found from simple counting.

The first person considered can be born on any day, which means the probability they will not be born on one of the 365 days of the year is 0.

$$P(1) = 1 - P(\text{not } 1) = 1 - 0 = 1 \quad (365/365)$$

The probability the second person will be born on the same day as the first person is 1 in 365. (The second person can be born on any other of the 364 days.) The probability that the second person is born on any other day is

$$P(2) = 1 - P(\text{not } 2) = 1 - \frac{1}{365} = \frac{364}{365}$$

The third person cannot have been born on either of the same days as the first and second people, which has a 2 in 365 probability of happening. The probability that the third person is born on any other day is

$$P(3) = 1 - P(\text{not } 3) = 1 - \frac{2}{365} = \frac{363}{365}$$

This logic continues to the seventh person. The probability that all seven conditions are simultaneously satisfied is

$$P(7 \text{ distinct birthdays})$$
$$= P(1) \times P(2) \times P(3) \times P(4) \times P(5)$$
$$\quad \times P(6) \times P(7)$$
$$= \left(\frac{365}{365}\right)\left(\frac{364}{365}\right)\left(\frac{363}{365}\right)\left(\frac{362}{365}\right)\left(\frac{361}{365}\right)$$
$$\quad \times \left(\frac{360}{365}\right)\left(\frac{359}{365}\right)$$
$$= 0.9438 \quad (0.94)$$

The answer is (D).

2. The mean is

$$\overline{X} = (1/n)\sum_{i=1}^{n} X_i$$
$$= \left(\frac{1}{12}\right)\begin{pmatrix} 3+4+4+5 \\ +8+8+8+10 \\ +11+15+18+20 \end{pmatrix}$$
$$= 9.5$$

The answer is (B).

3. The standard deviation is calculated using the sample mean as an unbiased estimator of the population mean.

$$\sigma = \sqrt{(1/N)\sum (X_i - \mu)^2} \approx \sqrt{(1/N)\sum (X_i - \overline{X})^2}$$
$$= \sqrt{\left(\frac{1}{8}\right)\begin{pmatrix} (2-7.25)^2 + (3-7.25)^2 \\ + (5-7.25)^2 + (8-7.25)^2 \\ + (8-7.25)^2 + (10-7.25)^2 \\ + (10-7.25)^2 + (12-7.25)^2 \end{pmatrix}}$$
$$= 3.34 \quad (3.3)$$

The answer is (C).

4. The sample standard deviation is

$$s = \sqrt{[1/(n-1)]\sum_{i=1}^{n}(X_i - \overline{X})^2}$$
$$= \sqrt{\left(\frac{1}{6-1}\right)\begin{pmatrix} (10-13)^2 + (12-13)^2 \\ + (13-13)^2 + (14-13)^2 \\ + (14-13)^2 + (15-13)^2 \end{pmatrix}}$$
$$= 1.79 \quad (1.8)$$

The answer is (D).

5. The variance is

$$\sigma^2 = (10.4)^2 = 108 \quad (110)$$

The answer is (C).

6. The sample variance is

$$s^2 = (4.2)^2 = 17.64 \quad (18)$$

The answer is (B).

7. There are 50 odd-numbered balls. Including ball 100, there are 20 balls with numbers greater than 80.

$$P(A) = P(\text{ball is odd}) = \frac{50}{100} = 0.5$$
$$P(B) = P(\text{ball} > 80) = \frac{20}{100} = 0.2$$

It is possible for the number on the selected ball to be both odd and greater than 80. Use the law of total probability.

$$P(A+B) = P(A) + P(B) - P(A,B)$$
$$= P(A) + P(B) - P(A)P(B)$$
$$P(\text{odd or} > 80) = 0.5 + 0.2 - (0.5)(0.2) = 0.6$$

The answer is (C).

8. Find the standard normal values for the two points of interest.

$$Z_{16\%} = \frac{x - \mu}{\sigma} = \frac{16\% - 14.2\%}{2.3\%}$$
$$= 0.78 \quad [\text{use } 0.80]$$

$$Z_{12\%} = \frac{x - \mu}{\sigma} = \frac{12\% - 14.2\%}{2.3\%}$$
$$= -0.96 \quad [\text{use } -1.00]$$

Use the unit normal distribution table. The probabilities being sought can be found from the values of $R(x)$ for both standard normal values. $R(0.80) = 0.2119$ and $R(1.00) = 0.1587$. The probability that the sample will fall outside these values is the sum of the two values.

$$P(x < 12\% \text{ or } x > 16\%) = 0.2119 + 0.1587$$
$$= 0.3706 \quad (0.37)$$

The answer is (C).

9. Find the probability of exactly 2 heads being thrown. The probability will be the quotient of the total number of possible combinations of six objects taken two at a time and the total number of possible outcomes from tossing six fair coins. The total number of possible outcomes is $(2)^6 = 64$. The total number of possible combinations in which exactly two heads are thrown is

$$C(n, r) = \frac{n!}{r!(n-r)!} = \frac{6!}{2!(6-2)!}$$
$$= 15$$

The probability of exactly two heads out of six fair coins is

$$P(A) = P(2 \text{ heads}) = \frac{15}{64} = 0.234$$

The probability of exactly three heads being thrown is found similarly. The total number of possible combinations in which exactly three heads are thrown is

$$C(n, r) = \frac{n!}{r!(n-r)!} = \frac{6!}{3!(6-3)!}$$
$$= 20$$

The probability of exactly three heads out of six fair coins is

$$P(B) = P(3 \text{ heads}) = \frac{20}{64} = 0.313$$

From the law of total probability, the probability that either of these outcomes will occur is the sum of the individual probabilities that the outcomes will occur, minus the probability that both will occur. These two outcomes are mutually exclusive (i.e., both cannot occur), so the probability of both happening is 0. The total probability is

$$P(2 \text{ heads or } 3 \text{ heads}) = P(A) + P(B) - P(A, B)$$
$$= 0.234 + 0.313 - 0$$
$$= 0.547 \quad (0.55)$$

The answer is (B).

10. If events A and B are mutually exclusive, the probability of both occurring is zero. However, either event could occur by itself, and the probability of that is non-zero.

The answer is (C).

11. For a fair die, the probability of any face turning up is $1/6$. There are two ways to win, and there are four ways to lose. The expected value is

$$E[X] = \sum_{k=1}^{n} x_k f(x_k) = (\$10)\left((2)\left(\frac{1}{6}\right)\right) + (-\$5)\left((4)\left(\frac{1}{6}\right)\right)$$
$$= \$0.00$$

The answer is (A).

12. A 95% confidence interval on mean transporter utilization means there is a 95% chance the population (or true) mean transporter utilization lies within the given interval.

The answer is (B).

13. Use the binomial probability function to calculate the probability that two of the seven samples will have been born on April 15. $x = 2$, and the sample size, n, is 7.

The probability that a person will have been born on April 15 is $1/365$. Therefore, the probability of "success," p, is $1/365$, and the probability of "failure," $q = 1 - p$, is $364/365$.

$$P_n(x) = \frac{n!}{x!(n-x)!} p^x q^{n-x}$$

$$P_7(2) = \left(\frac{7!}{2!(7-2)!}\right)\left(\frac{1}{365}\right)^2 \left(\frac{364}{365}\right)^{7-2}$$
$$= (21)\left(\frac{1}{365}\right)^2 \left(\frac{364}{365}\right)^5$$
$$= 1.555 \times 10^{-4} \quad (1.6 \times 10^{-4})$$

The answer is (D).

14. The arithmetic mean is

$$\overline{X} = (1/n)\sum_{i=1}^{n} X_i$$
$$= \left(\frac{1}{5}\right)(71.3 + 74.0 + 74.25 + 78.54 + 80.6)$$
$$= 75.738 \quad (75.7)$$

The sample standard deviation is

$$s = \sqrt{[1/(n-1)]\sum_{i=1}^{n}(X_i - \overline{X})^2}$$

$$= \sqrt{\left(\frac{1}{5-1}\right)\begin{pmatrix}(71.3 - 75.738)^2 + (74.0 - 75.738)^2 \\ + (74.25 - 75.738)^2 \\ + (78.54 - 75.738)^2 \\ + (80.6 - 75.738)^2\end{pmatrix}}$$

$$= 3.756 \quad (3.8)$$

The answer is (D).

15. The binomial probability function can be used to determine the probability of three heads in four trials.

$$p = P(\text{heads}) = 0.5$$
$$q = P(\text{not heads}) = 1 - 0.5 = 0.5$$
$$n = \text{number of trials} = 4$$
$$x = \text{number of successes} = 3$$

From the binomial function,

$$P_n(x) = \frac{n!}{x!(n-x)!}p^x q^{n-x}$$
$$= \left(\frac{4!}{3!(4-3)!}\right)(0.5)^3(0.5)^{4-3}$$
$$= 0.25 \quad (1/4)$$

The answer is (A).

16. The complement of set A contains all of the members of set Q that are not members of set A: $(5, 6, 8, 10)$.

The union of the complement of set A and set B is the set of all members appearing in either.

$$\overline{A} \cup B = (5, 6, 8, 10) \cup (4, 5, 9, 10)$$
$$= (4, 5, 6, 8, 9, 10)$$

The answer is (C).

17. Set "not A" consists of all universe elements not in set A: $(2, 4, 5, 7, 8)$.

The intersection of "not A" and B is the set of all elements appearing in both.

$$\overline{A} \cap B = (2, 4, 5, 7, 8) \cap (1, 2, 6, 7)$$
$$= (2, 7)$$

The answer is (A).

Fluid Properties

PRACTICE PROBLEMS

1. A leak from a faucet comes out in separate drops instead of a stream. The main cause of this phenomenon is

(A) gravity

(B) air resistance

(C) viscosity

(D) surface tension

2. A solid cylinder is concentric with a straight pipe. The cylinder is 0.5 m long and has an outside diameter of 8 cm. The pipe has an inside diameter of 8.5 cm. The annulus between the cylinder and the pipe contains stationary oil. The oil has a specific gravity of 0.92 and a kinematic viscosity of 5.57×10^{-4} m^2/s. The force needed to move the cylinder along the pipe at a constant velocity of 1 m/s is most nearly

(A) 5.9 N

(B) 12 N

(C) 26 N

(D) 55 N

3. Kinematic viscosity can be expressed in

(A) m^2/s

(B) s^2/m

(C) kg·s^2/m

(D) kg/s

4. Which three of the following must be satisfied by the flow of any fluid, whether real or ideal?

I. Newton's second law of motion

II. the continuity equation

III. uniform velocity distribution

IV. Newton's law of viscosity

V. conservation of energy

(A) I, II, and III

(B) I, II, and V

(C) I, III, and V

(D) II, IV, and V

5. 15 kg of a fluid with a density of 790 kg/m^3 is mixed with 10 kg of water. The volumes are additive, and the resulting mixture is homogeneous. The specific volume of the resulting mixture is most nearly

(A) 0.0012 m^3/kg

(B) 0.0027 m^3/kg

(C) 0.0047 m^3/kg

(D) 0.0061 m^3/kg

6. The rise or fall of liquid in a small-diameter capillary tube is NOT affected by

(A) adhesive forces

(B) cohesive forces

(C) surface tension

(D) fluid viscosity

7. A capillary tube 3.8 mm in diameter is placed in a beaker of 40°C distilled water. The surface tension is 0.0696 N/m, and the angle made by the water with the wetted tube wall is negligible. The specific weight of water at this temperature is 9.730 kN/m^3. The height to which the water will rise in the tube is most nearly

(A) 1.2 mm

(B) 3.6 mm

(C) 7.5 mm

(D) 9.2 mm

SOLUTIONS

1. Surface tension is caused by the molecular cohesive forces in a fluid. It is the main cause of the formation of drops of water.

The answer is (D).

2. Treat the cylinder as a moving plate, and use Newton's law of viscosity. Find the absolute viscosity of the oil.

$$\nu = \frac{\mu}{\rho}$$

$$\mu = \nu\rho = \left(5.57 \times 10^{-4}\ \frac{\text{m}^2}{\text{s}}\right)(0.92)\left(1000\ \frac{\text{kg}}{\text{m}^3}\right)$$

$$= 0.512\ \text{Pa·s}$$

The width of the separation between the cylinder and the pipe is

$$\delta = \frac{d_{\text{pipe}} - d_{\text{cylinder}}}{2} = \frac{8.5\ \text{cm} - 8\ \text{cm}}{2}$$

$$= 0.25\ \text{cm}$$

The interval surface area of the cylinder is

$$A = \pi d L = \pi \left(\frac{8\ \text{cm}}{100\ \frac{\text{cm}}{\text{m}}}\right)(0.5\ \text{m})$$

$$= 0.126\ \text{m}^2$$

Find the force needed.

$$\frac{F}{A} = \mu\left(\frac{d\text{v}}{dy}\right)$$

$$F = A\mu\left(\frac{d\text{v}}{dy}\right) \approx A\mu\left(\frac{\Delta\text{v}}{\Delta y}\right)$$

$$= (0.126\ \text{m}^2)(0.512\ \text{Pa·s})\left(\frac{1\ \frac{\text{m}}{\text{s}}}{0.25\ \text{cm}}\right)\left(100\ \frac{\text{cm}}{\text{m}}\right)$$

$$= 25.8\ \text{N} \quad (26\ \text{N})$$

The answer is (C).

3. Typical units of kinematic viscosity are m^2/s.

The answer is (A).

4. Newton's second law, the continuity equation, and the principle of conservation of energy always apply for any fluid.

The answer is (B).

5. Calculate the volumes. Use a standard water density of 1000 kg/m³.

$$\rho = \frac{m}{V}$$

$$V = \frac{m}{\rho}$$

$$V_{\text{water}} = \frac{m}{\rho} = \frac{10\ \text{kg}}{1000\ \frac{\text{kg}}{\text{m}^3}} = 0.010\ \text{m}^3$$

$$V_{\text{fluid}} = \frac{m}{\rho} = \frac{15\ \text{kg}}{790\ \frac{\text{kg}}{\text{m}^3}} = 0.019\ \text{m}^3$$

The total volume is

$$V_{\text{total}} = V_{\text{water}} + V_{\text{fluid}} = 0.010\ \text{m}^3 + 0.019\ \text{m}^3$$

$$= 0.029\ \text{m}^3$$

The density of the mixture is the total mass divided by the total volume.

$$\rho_{\text{mixture}} = \frac{m_{\text{water}} + m_{\text{fluid}}}{V_{\text{total}}} = \frac{10\ \text{kg} + 15\ \text{kg}}{0.029\ \text{m}^3}$$

$$= 862\ \text{kg/m}^3$$

The specific volume of the mixture is the reciprocal of its density.

$$v_{\text{mixture}} = \frac{1}{\rho_{\text{mixture}}} = \frac{1}{862\ \frac{\text{kg}}{\text{m}^3}}$$

$$= 0.00116\ \text{m}^3/\text{kg} \quad (0.0012\ \text{m}^3/\text{kg})$$

The answer is (A).

6. The height of capillary rise is

$$h = 4\sigma \cos\beta / \gamma d$$

σ is the surface tension of the fluid, β is the angle of contact, γ is the specific weight of the liquid, and d is the diameter of the tube.

The viscosity of the fluid is not directly relevant to the height of capillary rise.

The answer is (D).

7. Since the contact angle is neglible, use 0° for β. The capillary rise in liquid is

$$h = 4\sigma \cos\beta / \gamma d$$

$$= \frac{(4)\left(0.0696\ \frac{\text{N}}{\text{m}}\right)\cos 0° \left(1000\ \frac{\text{mm}}{\text{m}}\right)}{\left(9.730\ \frac{\text{kN}}{\text{m}^3}\right)\left(1000\ \frac{\text{N}}{\text{kN}}\right)(3.8\ \text{mm})}$$

$$= 7.53 \times 10^{-3}\ \text{m} \quad (7.5\ \text{mm})$$

The answer is (C).

8 Fluid Statics

PRACTICE PROBLEMS

1. A barometer contains mercury with a density of 13 600 kg/m³. Atmospheric conditions are 95.8 kPa and 20°C. At 20°C, the vapor pressure of the mercury is 0.000173 kPa. The column of mercury will rise to a height of most nearly

(A) 0.38 m

(B) 0.48 m

(C) 0.72 m

(D) 0.82 m

2. The manometer shown contains water, mercury, and glycerine. The specific gravity of mercury is 13.6, and the specific gravity of glycerine is 1.26.

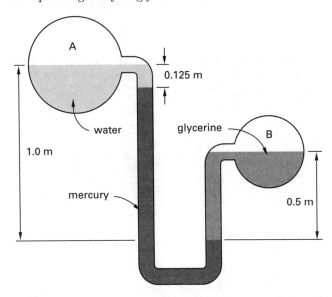

What is most nearly the difference in pressure between points A and B?

(A) 35 kPa

(B) 42 kPa

(C) 55 kPa

(D) 110 kPa

3. An open water manometer is used to measure the pressure in a tank. The tank is cylindrical with hemispherical ends. The tank is half-filled with 50 000 kg of a liquid chemical that is not miscible in water. The manometer tube is filled with liquid chemical up to the water.

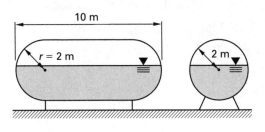

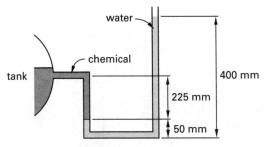

What is most nearly the pressure in the tank relative to the atmospheric pressure?

(A) 1.4 kPa

(B) 1.9 kPa

(C) 2.4 kPa

(D) 3.4 kPa

4. A pressure vessel is connected to a simple U-tube open to the atmosphere as shown. A 10 cm deflection of mercury is observed. The density of mercury is 13 600 kg/m³. The atmospheric pressure is 101 kPa.

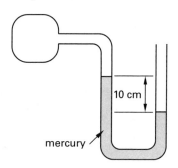

The vacuum within the vessel is most nearly

(A) 1.0 kPa

(B) 13 kPa

(C) 39 kPa

(D) 78 kPa

5. A tank contains a gate 2 m tall and 5 m wide as shown. The tank is filled with water to a depth of 10 m.

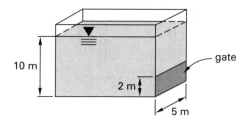

The total force on the gate is most nearly

(A) 90 kN

(B) 440 kN

(C) 880 kN

(D) 980 kN

6. A 1 m × 2 m inclined plate is submerged as shown.

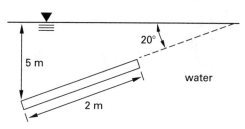

The normal force acting on the upper surface of the plate is most nearly

(A) 32 kN

(B) 56 kN

(C) 68 kN

(D) 91 kN

7. The water tank shown has a width of 0.3 m. The rounded corner has a radius of 0.9 m.

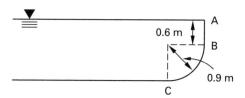

The magnitude and direction (in degrees from the horizontal) of the water force over the length of wall from point A to point C is most nearly

(A) 3400 N, 46°

(B) 3400 N, 73°

(C) 4800 N, 46°

(D) 4800 N, 73°

SOLUTIONS

1. Find the height of the mercury in the column.

$$p_A = p_B + \rho g h$$

$$h = \frac{p_A - p_B}{\rho g}$$

$$= \frac{(95.8 \text{ kPa} - 0.000173 \text{ kPa})\left(1000 \frac{\text{Pa}}{\text{kPa}}\right)}{\left(13\,600 \frac{\text{kg}}{\text{m}^3}\right)\left(9.81 \frac{\text{m}}{\text{s}^2}\right)}$$

$$= 0.7181 \text{ m} \quad (0.72 \text{ m})$$

The answer is (C).

2. The manometer can be labeled as shown.

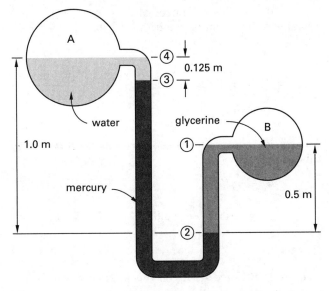

The pressure at level 2 is the same in both (left and right) legs of the manometer. For the left leg,

$$p_2 = p_A + \rho_{\text{water}} g h_{3\text{-}4} + \rho_{\text{mercury}} g h_{2\text{-}3}$$

For the right leg,

$$p_2 = p_B + \rho_{\text{glycerine}} g h_{1\text{-}2}$$

Equating these two equations for p_2 and solving for the pressure difference $p_A - p_B$ gives

$$p_A - p_B = g(\rho_{\text{glycerine}} h_{1\text{-}2} - \rho_{\text{water}} h_{3\text{-}4} - \rho_{\text{Hg}} h_{2\text{-}3})$$

$$= g\rho_{\text{water}} \begin{pmatrix} SG_{\text{glycerine}} h_{1\text{-}2} - SG_{\text{water}} h_{3\text{-}4} \\ - SG_{\text{Hg}} h_{2\text{-}3} \end{pmatrix}$$

$$= \left(9.81 \frac{\text{m}}{\text{s}^2}\right)\left(1000 \frac{\text{kg}}{\text{m}^3}\right)$$

$$\times \begin{pmatrix} (1.26)(0.5 \text{ m}) - (1.00)(0.125 \text{ m}) \\ - (13.6)(1.0 \text{ m} - 0.125 \text{ m}) \end{pmatrix}$$

$$= -111\,785 \text{ Pa} \quad (110 \text{ kPa})$$

The answer is (D).

3. Calculate the density of the chemical from the volume and mass. The total volume of the tank is

$$V = \tfrac{4}{3}\pi r^3 + \pi r^2(L - 2r)$$

$$= \tfrac{4}{3}\pi(2 \text{ m})^3 + \pi(2 \text{ m})^2(10 \text{ m} - (2)(2 \text{ m}))$$

$$= 108.9 \text{ m}^3$$

The contents have a mass of 50 000 kg, and the tank is half full, so the density of the chemical is

$$\rho_{\text{chemical}} = \frac{m}{V} = \frac{50\,000 \text{ kg}}{\left(\tfrac{1}{2}\right)(108.9 \text{ m}^3)}$$

$$= 918.2 \text{ kg/m}^3$$

The relative pressure is

$$p_0 - p_2 = \rho_{\text{water}} g h_2 - \rho_{\text{chemical}} g h_1$$

$$= \left(1000 \frac{\text{kg}}{\text{m}^3}\right)\left(9.81 \frac{\text{m}}{\text{s}^2}\right)\left(\frac{400 \text{ mm} - 50 \text{ mm}}{1000 \frac{\text{mm}}{\text{m}}}\right)$$

$$- \left(918.2 \frac{\text{kg}}{\text{m}^3}\right)\left(9.81 \frac{\text{m}}{\text{s}^2}\right)\left(\frac{225 \text{ m}}{1000 \frac{\text{mm}}{\text{m}}}\right)$$

$$= 1407 \text{ Pa} \quad (1.4 \text{ kPa})$$

The answer is (A).

4. Vacuum is the difference between the atmospheric pressure and the absolute pressure in the vessel (i.e., is the gage pressure). The vacuum is

$$p_{\text{gage}} = -\rho g h$$
$$= -\left(13\,600\ \frac{\text{kg}}{\text{m}^3}\right)\left(9.81\ \frac{\text{m}}{\text{s}^2}\right)\left(\frac{10\ \text{cm}}{100\ \frac{\text{cm}}{\text{m}}}\right)$$
$$= -13\,342\ \text{Pa}\quad(13\ \text{kPa})$$

The answer is (B).

5. $h_1 = 10\ \text{m} - 2\ \text{m} = 8\ \text{m}$. The average pressure is

$$\overline{p} = \tfrac{1}{2}\rho g(h_1 + h_2)$$
$$= \left(\tfrac{1}{2}\right)\left(1000\ \frac{\text{kg}}{\text{m}^3}\right)\left(9.81\ \frac{\text{m}}{\text{s}^2}\right)(8\ \text{m} + 10\ \text{m})$$
$$= 88\,290\ \text{Pa}$$

The total force acting on the gate is

$$R = \overline{p}A = (88\,290\ \text{Pa})\big((2\ \text{m})(5\ \text{m})\big)$$
$$= 882\,900\ \text{N}\quad(880\ \text{kN})$$

The answer is (C).

6. The upper edge of the plate is at a depth of

$$h_1 = 5\ \text{m} - (2\ \text{m})\sin 20° = 4.32\ \text{m}$$

The average pressure is

$$\overline{p} = \tfrac{1}{2}\rho g(h_1 + h_2)$$
$$= \left(\tfrac{1}{2}\right)\left(1000\ \frac{\text{kg}}{\text{m}^3}\right)\left(9.81\ \frac{\text{m}}{\text{s}^2}\right)(4.32\ \text{m} + 5\ \text{m})$$
$$= 45\,695\ \text{Pa}$$

The normal force acting on the plate is

$$R = \overline{p}A = (45\,695\ \text{Pa})\big((1\ \text{m})(2\ \text{m})\big)$$
$$= 91\,390\ \text{N}\quad(91\ \text{kN})$$

The answer is (D).

7. Find separately the horizontal and vertical components of the force acting on the wall from point A to point C. For the horizontal component, $h_1 = 0\ \text{m}$, and $h_2 = 0.6\ \text{m} + 0.9\ \text{m} = 1.5\ \text{m}$.

$$\overline{p}_x = \tfrac{1}{2}\rho g(h_1 + h_2)$$
$$= \left(\tfrac{1}{2}\right)\left(1000\ \frac{\text{kg}}{\text{m}^3}\right)\left(9.81\ \frac{\text{m}}{\text{s}^2}\right)(0\ \text{m} + 1.5\ \text{m})$$
$$= 7358\ \text{Pa}$$

The horizontal component of the force is

$$R_x = \overline{p}_x A = (7358\ \text{Pa})\big((1.5\ \text{m})(0.3\ \text{m})\big)$$
$$= 3311\ \text{N}$$

For the vertical component, calculate the weight of the water above the section of wall from point B to point C. The volume consists of rectangular prism $0.3\ \text{m} \times 0.6\ \text{m} \times 0.9\ \text{m}$ plus a quarter of a cylinder, which has a radius of 0.9 m and a length of 0.3 m.

$$V = V_1 + V_2$$
$$= (0.3\ \text{m})(0.6\ \text{m})(0.9\ \text{m}) + \frac{\pi(0.9\ \text{m})^2(0.3\ \text{m})}{4}$$
$$= 0.3529\ \text{m}^3$$

The vertical component of the force equals the weight of this volume of water.

$$R_y = \rho g V$$
$$= \left(1000\ \frac{\text{kg}}{\text{m}^3}\right)\left(9.81\ \frac{\text{m}}{\text{s}^2}\right)(0.3529\ \text{m}^3)$$
$$= 3461\ \text{N}$$

The resultant force acting on this section of wall is

$$R = \sqrt{R_x^2 + R_y^2} = \sqrt{(3311\ \text{N})^2 + (3461\ \text{N})^2}$$
$$= 4790\ \text{N}\quad(4800\ \text{N})$$

The direction of the resultant force from the horizontal is

$$\theta = \arctan\frac{R_y}{R_x} = \arctan\frac{3461\ \text{N}}{3311\ \text{N}}$$
$$= 46.27°\quad(46°)$$

The answer is (C).

9 Fluid Dynamics

PRACTICE PROBLEMS

1. 2750 kg/min of water are pumped between two reservoirs through 55 m of 150 mm inside diameter pipe. The pipe has a Darcy friction factor of 0.02. The water has a density of 1000 kg/m^3. The friction head loss over the entire length of the pipe is most nearly

(A) 0.96 m

(B) 1.7 m

(C) 2.1 m

(D) 2.5 m

2. Minor losses are decreases in pressure due to friction

(A) in fully developed turbulent flow through pipes of constant area

(B) in valves, tees, and elbows

(C) that can usually be ignored

(D) in fully developed turbulent flow in nonconstant area pipes

3. Water flows at constant velocity through a horizontal pipe. The gage pressure at point 1 is 1.03 kPa, and the gage pressure at point 2 downstream is 1.00 kPa. The head loss between points 1 and 2 is most nearly

(A) 3.1×10^{-3} m

(B) 2.3×10^{-2} m

(C) 3.1×10^{-2} m

(D) 2.3 m

4. The hydraulic radius of a fluid conduit is the

(A) mean radius from the center of flow to the wetted side of the conduit

(B) cross-sectional area of the conduit divided by the wetted perimeter

(C) wetted perimeter of the conduit divided by the area of flow

(D) cross-sectional area in flow divided by the wetted perimeter

5. Water flows through a multisectional pipe placed horizontally on the ground. The velocity is 3.0 m/s at the entrance and 2.1 m/s at the exit. If friction is neglected, the pressure difference between these two points is most nearly

(A) 0.20 kPa

(B) 2.3 kPa

(C) 28 kPa

(D) 110 kPa

6. Fluid flows at 5 m/s through a section of 5 cm diameter pipe. This section is connected to a section of 10 cm diameter pipe. The velocity of the fluid in the 10 cm diameter section is most nearly

(A) 1.0 m/s

(B) 1.3 m/s

(C) 2.5 m/s

(D) 10 m/s

7. An open channel has a cross-sectional area of flow of 0.5 m^2, a hydraulic radius of 0.15 m, and a roughness coefficient of 0.15. The slope of the hydraulic gradient needed to achieve a flow rate of 10 L/s is most nearly

(A) 1.1×10^{-4}

(B) 6.7×10^{-4}

(C) 1.1×10^{-3}

(D) 6.7×10^{-3}

8. A pipe carrying an incompressible fluid has a diameter of 100 mm at point 1 and a diameter of 50 mm at point 2. The velocity of the fluid at point 1 is 0.3 m/s. What is most nearly the velocity at point 2?

(A) 0.95 m/s

(B) 1.2 m/s

(C) 2.1 m/s

(D) 3.5 m/s

9. Water flows through a converging fitting as shown and discharges freely to the atmosphere at the exit. Flow is incompressible, and friction is negligible.

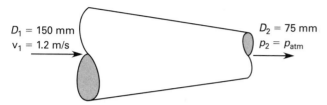

The gage pressure at the inlet is most nearly

(A) 10.2 kPa

(B) 10.8 kPa

(C) 11.3 kPa

(D) 12.7 kPa

10. A 90° reducing elbow connects a vertical and horizontal water pipe. Water discharges to the right, as shown.

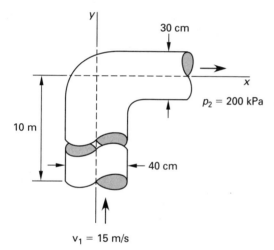

If the mass of the fluid is neglected, the horizontal force needed to hold the reducer elbow in a stationary position is most nearly

(A) 24 kN to the right

(B) 57 kN to the left

(C) 64 kN to the right

(D) 71 kN to the left

11. The basis of the Bernoulli equation for fluid flow is the

(A) principle of conservation of mass

(B) principle of conservation of energy

(C) continuity equation

(D) principle of conservation of momentum

12. The open channel shown is constructed of smooth concrete. It has a slope of 0.005, and the Hazen-Williams roughness coefficient is 130.

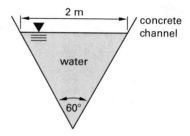

If the fluid depth is reduced to half of what it is now, the flow rate will be most nearly

(A) 0.80 m³/s

(B) 1.0 m³/s

(C) 1.5 m³/s

(D) 2.2 m³/s

SOLUTIONS

1. The flow area is

$$A = \frac{\pi D^2}{4} = \frac{\pi (150 \text{ mm})^2}{(4)\left(1000 \frac{\text{mm}}{\text{m}}\right)^2} = 0.0177 \text{ m}^2$$

The flow rate is

$$Q = \frac{\dot{m}}{\rho} = \frac{2750 \frac{\text{kg}}{\text{min}}}{\left(1000 \frac{\text{kg}}{\text{m}^3}\right)\left(60 \frac{\text{s}}{\text{min}}\right)} = 0.0458 \text{ m}^3/\text{s}$$

The flow velocity is

$$\text{v} = \frac{Q}{A} = \frac{0.0458 \frac{\text{m}^3}{\text{s}}}{0.0177 \text{ m}^2} = 2.59 \text{ m/s}$$

Use the Darcy equation to find the friction loss.

$$h_f = f \frac{L}{D} \frac{\text{v}^2}{2g}$$

$$= (0.02) \left(\frac{(55 \text{ m})\left(1000 \frac{\text{mm}}{\text{m}}\right)}{150 \text{ mm}}\right) \left(\frac{\left(2.59 \frac{\text{m}}{\text{s}}\right)^2}{(2)\left(9.81 \frac{\text{m}}{\text{s}^2}\right)}\right)$$

$$= 2.51 \text{ m} \quad (2.5 \text{ m})$$

The answer is (D).

2. Minor losses are friction losses caused by fittings, changes in direction, and changes in flow area. The flow regime (laminar or turbulent) and the pipe cross section are irrelevant.

The answer is (B).

3. From the energy conservation equation,

$$\frac{p_1}{\rho g} + z_1 + \frac{\text{v}_1^2}{2g} = \frac{p_2}{\rho g} + z_2 + \frac{\text{v}_2^2}{2g} + h_f$$

$$z_1 = z_2$$

$$\text{v}_1 = \text{v}_2$$

$$\frac{p_1}{\rho g} = \frac{p_2}{\rho g} + h_f$$

$$h_f = \frac{p_1 - p_2}{\rho g}$$

$$= \frac{(1.03 \text{ kPa} - 1.00 \text{ kPa})\left(1000 \frac{\text{Pa}}{\text{kPa}}\right)}{\left(1000 \frac{\text{kg}}{\text{m}^3}\right)\left(9.81 \frac{\text{m}}{\text{s}^2}\right)}$$

$$= 3.1 \times 10^{-3} \text{ m}$$

The answer is (A).

4. The hydraulic radius is defined as the cross-sectional area in flow divided by the wetted perimeter.

$$R_H = \frac{\text{cross-sectional area}}{\text{wetted perimeter}}$$

The answer is (D).

5. From the Bernoulli equation,

$$\frac{p_2}{\rho g} + \frac{\text{v}_2^2}{2g} + z_2 = \frac{p_1}{\rho g} + \frac{\text{v}_1^2}{2g} + z_1$$

$$z_2 = z_1 \quad \text{[since the pipe is horizontal]}$$

$$\Delta p = p_2 - p_1$$

$$= \rho g \left(\frac{\text{v}_1^2 - \text{v}_2^2}{2g}\right)$$

$$= \left(\frac{\rho}{2}\right)(\text{v}_1^2 - \text{v}_2^2)$$

$$= \left(\frac{1000 \frac{\text{kg}}{\text{m}^3}}{2}\right)\left(\left(3.0 \frac{\text{m}}{\text{s}}\right)^2 - \left(2.1 \frac{\text{m}}{\text{s}}\right)^2\right)$$

$$= 2295 \text{ Pa} \quad (2.3 \text{ kPa})$$

The answer is (B).

6. Q is constant.

$$\text{v}_1 A_1 = \text{v}_2 A_2$$

$$\text{v}_2 = \frac{\text{v}_1 A_1}{A_2} = \frac{\text{v}_1 D_1^2}{D_2^2}$$

$$= \frac{\left(5 \frac{\text{m}}{\text{s}}\right)(5 \text{ cm})^2 \left(100 \frac{\text{cm}}{\text{m}}\right)^2}{(10 \text{ cm})^2 \left(100 \frac{\text{cm}}{\text{m}}\right)^2}$$

$$= 1.25 \text{ m/s} \quad (1.3 \text{ m/s})$$

The answer is (B).

7. The volumetric flow rate is

$$Q = \frac{10 \frac{\text{L}}{\text{s}}}{1000 \frac{\text{L}}{\text{m}^3}} = 0.01 \text{ m}^3/\text{s}$$

The velocity needed is

$$\text{v} = \frac{Q}{A} = \frac{0.01 \frac{\text{m}^3}{\text{s}}}{0.5 \text{ m}^2} = 0.02 \text{ m/s}$$

Use Manning's equation to find the slope needed to achieve this velocity.

$$v = (K/n) R_H^{2/3} S^{1/2}$$

$$S = \left(\frac{vn}{KR_H^{2/3}}\right)^2$$

$$= \left(\frac{\left(0.02 \frac{\text{m}}{\text{s}}\right)(0.15)}{(1.0)(0.15 \text{ m})^{2/3}}\right)^2$$

$$= 0.0001129 \quad (1.1 \times 10^{-4})$$

The answer is (A).

8. Use the continuity equation for an incompressible fluid and solve for the velocity at point 2.

$$A_1 v_1 = A_2 v_2$$

$$\left(\frac{\pi D_1^2}{4}\right) v_1 = \left(\frac{\pi D_2^2}{4}\right) v_2$$

$$v_2 = \left(\frac{D_1}{D_2}\right)^2 v_1 = \left(\frac{100 \text{ mm}}{50 \text{ mm}}\right)^2 \left(0.3 \frac{\text{m}}{\text{s}}\right)$$

$$= 1.2 \text{ m/s}$$

The answer is (B).

9. From the continuity equation for incompressible flow,

$$A_1 v_1 = A_2 v_2$$

$$v_2 = \frac{A_1 v_1}{A_2} = \left(\frac{D_1}{D_2}\right)^2 v_1$$

$$= \left(\frac{150 \text{ mm}}{75 \text{ mm}}\right)^2 \left(1.2 \frac{\text{m}}{\text{s}}\right)$$

$$= 4.8 \text{ m/s}$$

Use the Bernoulli equation.

$$\frac{p_2}{\rho} + \frac{v_2^2}{2} + z_2 g = \frac{p_1}{\rho} + \frac{v_1^2}{2} + z_1 g$$

$$z_1 = z_2$$

$$p_2 = 0 \quad [\text{gage}]$$

$$p_1 = \frac{\rho}{2}(v_2^2 - v_1^2)$$

$$= \left(\frac{1000 \frac{\text{kg}}{\text{m}^3}}{2}\right)\left(\left(4.8 \frac{\text{m}}{\text{s}}\right)^2 - \left(1.2 \frac{\text{m}}{\text{s}}\right)^2\right)$$

$$= 10\,800 \text{ Pa} \quad (10.8 \text{ kPa})$$

The answer is (B).

10. The free-body diagram of the fluid control volume in the reducer is

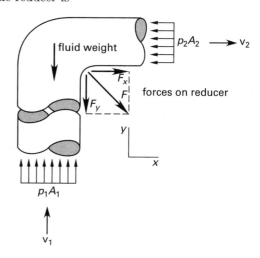

The pipe areas are

$$A_1 = \frac{\pi D_1^2}{4} = \frac{\pi (40 \text{ cm})^2}{(4)\left(100 \frac{\text{cm}}{\text{m}}\right)^2} = 0.1257 \text{ m}^2$$

$$A_2 = \frac{\pi D_2^2}{4} = \frac{\pi (30 \text{ cm})^2}{(4)\left(100 \frac{\text{cm}}{\text{m}}\right)^2} = 0.0707 \text{ m}^2$$

From the continuity equation,

$$Q = A_1 v_1 = A_2 v_2$$

$$v_2 = \frac{A_1 v_1}{A_2} = \frac{(0.1257 \text{ m}^2)\left(15 \frac{\text{m}}{\text{s}}\right)}{0.0707 \text{ m}^2}$$

$$= 26.67 \text{ m/s}$$

The horizontal force on the elbow comes from the horizontal pressure and the change in fluid momentum.

$$F_x = (p_2 A_2 + Q \rho v_2) \sin \alpha$$

Since $\alpha = 90°$ and $\sin 90° = 1$,

$$F_x = p_2 A_2 + Q \rho v_2$$
$$= p_2 A_2 + A_2 \rho v_2^2$$
$$= (200 \text{ kPa})(0.0707 \text{ m}^2)$$
$$+ \frac{(0.0707 \text{ m}^2)\left(1000 \frac{\text{kg}}{\text{m}^3}\right)\left(26.67 \frac{\text{m}}{\text{s}}\right)^2}{1000 \frac{\text{N}}{\text{kN}}}$$

$$= 64.40 \text{ kN} \quad [\text{to the right}]$$

The force exerted by the fluid on the reducer is equal and opposite to this force, so the x-component of the resultant force on the reducer is $F_x = -64.40$ kN to the left. The horizontal force, F_x, required to hold the reducer in a stationary position is $F_x = 64.40$ kN (64 kN) to the right.

The answer is (C).

11. The Bernoulli equation is derived from the principle of conservation of energy.

The answer is (B).

12. The flow has the cross-sectional shape of an equilateral triangle. When the depth of flow is halved, the triangle will have three sides of 1 m. The new depth of flow will be

$$d = \frac{\sqrt{3} \text{ m}}{2}$$

The area in flow will be

$$A = (2)(\tfrac{1}{2})bd = bd$$
$$= (\tfrac{1}{2} \text{ m})\left(\frac{\sqrt{3}}{2} \text{ m}\right)$$
$$= \frac{\sqrt{3}}{4} \text{ m}^2$$

The wetted perimeter is

$$\text{wetted perimeter} = 1 \text{ m} + 1 \text{ m} = 2 \text{ m}$$

The hydraulic radius is

$$R_H = \frac{\text{cross-sectional area}}{\text{wetted perimeter}} = \frac{\frac{\sqrt{3}}{4} \text{ m}^2}{2 \text{ m}} = \frac{\sqrt{3}}{8} \text{ m}$$

Use the Hazen-Williams equation to find the new velocity of flow.

$$\text{v} = k_1 C R_H^{0.63} S^{0.54}$$
$$= (0.849)(130)\left(\frac{\sqrt{3}}{8} \text{ m}\right)^{0.63}(0.005)^{0.54}$$
$$= 2.408 \text{ m/s}$$

The rate of flow will be

$$Q = A\text{v} = \left(\frac{\sqrt{3}}{4} \text{ m}^2\right)\left(2.408 \ \frac{\text{m}}{\text{s}}\right)$$
$$= 1.04 \text{ m}^3/\text{s} \quad (1.0 \text{ m}^3/\text{s})$$

The answer is (B).

10 Fluid Measurement and Similitude

PRACTICE PROBLEMS

1. A pitot tube is used to measure the flow of an incompressible fluid with a density of 926 kg/m³. The velocity is measured as 2 m/s, and the stagnation pressure is 14.1 kPa. The static pressure of the fluid where the measurement is taken is most nearly

(A) 10.4 kPa

(B) 11.7 kPa

(C) 12.2 kPa

(D) 13.5 kPa

2. A sharp-edged orifice with a 50 mm diameter opening in the vertical side of a large tank discharges under a head of 5 m. The coefficient of contraction is 0.62, and the coefficient of velocity is 0.98.

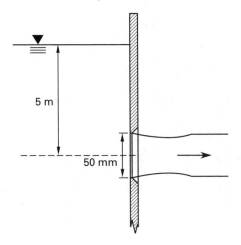

The rate of discharge is most nearly

(A) 0.00031 m³/s

(B) 0.0040 m³/s

(C) 0.010 m³/s

(D) 0.012 m³/s

3. The velocity of the water in the stream shown is 1.2 m/s.

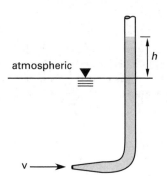

The height of water in the pitot tube is most nearly

(A) 3.7 cm

(B) 4.6 cm

(C) 7.3 cm

(D) 9.2 cm

4. A horizontal venturi meter with a diameter of 15 cm at the throat is installed in a 45 cm water main. A differential manometer gauge is partly filled with mercury (the remainder of the tube is filled with water) and connected to the meter at the throat and inlet. The mercury column stands 37.5 cm higher in one leg than in the other. The specific gravity of mercury is 13.6.

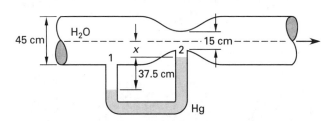

Neglecting friction, the flow through the meter is most nearly

(A) 0.10 m³/s

(B) 0.17 m³/s

(C) 0.23 m³/s

(D) 0.28 m³/s

5. A 1:1 model of a torpedo is tested in a wind tunnel according to the Reynolds number similarity. At the testing temperature, the kinematic viscosity of air is 1.41×10^{-5} m^2/s, and the kinematic viscosity of water is 1.31×10^{-6} m^2/s. If the velocity of the torpedo in water is 7 m/s, the air velocity in the wind tunnel should be most nearly

(A) 0.62 m/s
(B) 7.0 m/s
(C) 18 m/s
(D) 75 m/s

6. A 2 m tall, 0.5 m inside diameter tank is filled with water. A 10 cm hole is opened 0.75 m from the bottom of the tank. Ignoring all orifice losses, the velocity of the exiting water is most nearly

(A) 4.75 m/s
(B) 4.80 m/s
(C) 4.85 m/s
(D) 4.95 m/s

7. Water flows from one reservoir to another through a perfectly insulated pipe. Between the two reservoirs, 100 m of head is lost due to friction. Water has a specific heat of 4180 J/kg·K. The increase in water temperature between the reservoirs is most nearly

(A) 0.23°C
(B) 0.52°C
(C) 0.70°C
(D) 1.0°C

SOLUTIONS

1. Solve the equation for velocity in a pitot tube for the static pressure.

$$v = \sqrt{\left(\frac{2}{\rho}\right)(p_0 - p_s)}$$

$$p_s = p_0 - \frac{\rho v^2}{2}$$

$$= 14.1 \text{ kPa} - \frac{\left(926 \frac{\text{kg}}{\text{m}^3}\right)\left(2 \frac{\text{m}}{\text{s}}\right)^2}{(2)\left(1000 \frac{\text{Pa}}{\text{kPa}}\right)}$$

$$= 12.2 \text{ kPa}$$

The answer is (C).

2. The area of the opening is

$$A = \frac{\pi D^2}{4} = \frac{\pi (50 \text{ mm})^2}{(4)\left(1000 \frac{\text{mm}}{\text{m}}\right)^2}$$

$$= 0.00196 \text{ m}^2$$

The coefficient of discharge is

$$C = C_c C_v = (0.62)(0.98)$$
$$= 0.6076$$

The discharge rate is

$$Q = CA_0\sqrt{2gh}$$

$$= (0.6076)(0.00196 \text{ m}^2)\sqrt{(2)\left(9.81 \frac{\text{m}}{\text{s}^2}\right)(5 \text{ m})}$$

$$= 0.012 \text{ m}^3/\text{s}$$

The answer is (D).

3. The difference in height between the pitot tube and the free-water surface is a measure of the difference in static and stagnation pressures. Solve for the height of the water.

$$v = \sqrt{\left(\frac{2}{\rho}\right)(p_0 - p_s)} = \sqrt{\left(\frac{2}{\rho}\right)\rho gh}$$

$$= \sqrt{2gh}$$

$$h = \frac{v^2}{2g} = \frac{\left(1.2 \frac{\text{m}}{\text{s}}\right)^2}{(2)\left(9.81 \frac{\text{m}}{\text{s}^2}\right)}$$

$$= 0.073 \text{ m} \quad (7.3 \text{ cm})$$

The answer is (C).

4. The areas of the pipes are

$$A_1 = \frac{\pi D^2}{4} = \frac{\pi (45 \text{ cm})^2}{(4)\left(100 \, \frac{\text{cm}}{\text{m}}\right)^2} = 0.159 \text{ m}^2$$

$$A_2 = \frac{\pi D^2}{4} = \frac{\pi (15 \text{ cm})^2}{(4)\left(100 \, \frac{\text{cm}}{\text{m}}\right)^2} = 0.0177 \text{ m}^2$$

The equation for flow through a venturi meter can be written in terms of a manometer fluid reading. For horizontal flow, $z_1 = z_2$.

$$Q = \frac{C_v A_2}{\sqrt{1 - \left(\frac{A_2}{A_1}\right)^2}} \sqrt{2g\left(\frac{p_1}{\gamma} + z_1 - \frac{p_2}{\gamma} - z_2\right)}$$

$$= \left(\frac{C_v A_2}{\sqrt{1 - \left(\frac{A_2}{A_1}\right)^2}}\right) \sqrt{\frac{2g(\rho_m - \rho)h}{\rho}}$$

Because friction is to be neglected, $C_v = 1$. (For venturi meters, C_v is usually very close to one because the diameter changes are gradual and there is little friction loss.)

$$Q = \left(\frac{C_v A_2}{\sqrt{1 - \left(\frac{A_2}{A_1}\right)^2}}\right) \sqrt{\frac{2g(\rho_m - \rho)h}{\rho}}$$

$$= \left(\frac{(1)(0.0177 \text{ m}^2)}{\sqrt{1 - \left(\frac{0.0177 \text{ m}^2}{0.159 \text{ m}^2}\right)^2}}\right)$$

$$\times \sqrt{\frac{(2)\left(9.81 \, \frac{\text{m}}{\text{s}^2}\right)\left(1000 \, \frac{\text{kg}}{\text{m}^3}\right) \times (13.6 - 1)(37.5 \text{ cm})}{\left(1000 \, \frac{\text{kg}}{\text{m}^3}\right)\left(100 \, \frac{\text{cm}}{\text{m}}\right)}}$$

$$= 0.171 \text{ m}^3/\text{s} \quad (0.17 \text{ m}^3/\text{s})$$

The answer is (B).

5. From the Reynolds number similarity,

$$\left[\frac{F_I}{F_V}\right]_p = \left[\frac{F_I}{F_V}\right]_m = \left[\frac{\text{v}l\rho}{\mu}\right]_p = \left[\frac{\text{v}l\rho}{\mu}\right]_m = [\text{Re}]_p = [\text{Re}]_m$$

The scale is 1:1, so the lengths of the prototype and model are the same ($l_m = l_p$).

The similarity equation reduces to

$$\left(\frac{\text{v}\rho}{\mu}\right)_p = \left(\frac{\text{v}\rho}{\mu}\right)_m$$

$$\left(\frac{\text{v}}{\nu}\right)_p = \left(\frac{\text{v}}{\nu}\right)_m$$

$$\text{v}_m = \text{v}_p \left(\frac{\nu_m}{\nu_p}\right) = \left(7 \, \frac{\text{m}}{\text{s}}\right) \left(\frac{1.41 \times 10^{-5} \, \frac{\text{m}^2}{\text{s}}}{1.31 \times 10^{-6} \, \frac{\text{m}^2}{\text{s}}}\right)$$

$$= 75.3 \text{ m/s} \quad (75 \text{ m/s})$$

The answer is (D).

6. The hydraulic head at the hole is

$$h = 2 \text{ m} - 0.75 \text{ m} = 1.25 \text{ m}$$

For an orifice discharging freely into the atmosphere,

$$Q = CA_0\sqrt{2gh}$$

As orifice losses are neglected, $C = 1$. Dividing both sides by A_0 gives

$$\text{v} = C\sqrt{2gh} = 1\sqrt{(2)\left(9.81 \, \frac{\text{m}}{\text{s}^2}\right)(1.25 \text{ m})}$$

$$= 4.95 \text{ m/s}$$

The answer is (D).

7. Convert the frictional head loss to specific energy loss.

$$\Delta E = h_f g = (100 \text{ m})\left(9.81 \, \frac{\text{m}}{\text{s}^2}\right)$$

$$= 981 \text{ m}^2/\text{s}^2 \quad (981 \text{ J/kg})$$

The temperature increase is

$$\Delta T = \frac{\Delta E}{c_p} = \frac{981 \, \frac{\text{J}}{\text{kg}}}{4180 \, \frac{\text{J}}{\text{kg}\cdot\text{K}}} = 0.2347 \text{K} \quad (0.23°\text{C})$$

The temperature difference in kelvins is the same as in degrees Celsius.

The answer is (A).

11 Compressible Fluid Dynamics

PRACTICE PROBLEMS

1. An aircraft flies through 20°C air at 1700 km/h. The molecular weight of air is 29 g/mol, and its ratio of specific heats is 1.4. What is most nearly the aircraft's Mach number?

(A) 0.98
(B) 1.4
(C) 1.9
(D) 5.3

2. At a point along the centerline of an isentropic nozzle, air is flowing at 300 m/s. The static temperature at that point is 40°C. The specific gas constant of air is 287 J/kg·K, and the ratio of specific heats is 1.4. What is most nearly the static air temperature at another point along the centerline of the nozzle where the Mach number is 2?

(A) −74°C
(B) −20°C
(C) 45°C
(D) 84°C

3. At a particular point in an air turbine, the pressure is 135 Pa, and the temperature is 440K. Air behaves as an ideal gas with a specific gas constant of 287 J/kg·K. What is most nearly the specific volume of the air at that point?

(A) 710 m³/kg
(B) 830 m³/kg
(C) 940 m³/kg
(D) 1100 m³/kg

4. At a particular point in a wind tunnel, supersonic air has a static temperature of 320K. The specific gas constant of air is 287 J/kg·K, and the ratio of specific heats is 1.4. If the Mach number at that point is 1.5, the velocity of air is most nearly

(A) 400 m/s
(B) 450 m/s
(C) 480 m/s
(D) 540 m/s

5. For a particular gas, the acoustic velocity varies only with which of the following parameters?

(A) specific gas constant
(B) ratio of specific heats
(C) channel wall friction
(D) absolute temperature

6. An aircraft flies at 26,000 ft above mean sea level where the temperature of the air is −30°C. The aircraft's speed is Mach 1.5. Most nearly, what is the Mach number just behind a normal shock attached to the aircraft's nose?

(A) 0.6
(B) 0.7
(C) 0.8
(D) 0.9

11-2　FE MECHANICAL PRACTICE PROBLEMS

SOLUTIONS

1. The molecular weight is

$$\text{MW}_{\text{air}} = \frac{\left(29 \frac{\text{g}}{\text{mol}}\right)\left(1000 \frac{\text{mol}}{\text{kmol}}\right)}{1000 \frac{\text{g}}{\text{kg}}} = 29 \text{ kg/kmol}$$

The speed of sound in the air is

$$c = \sqrt{kRT} = \sqrt{\frac{k\overline{R}T}{\text{MW}_{\text{air}}}}$$

$$= \sqrt{\frac{(1.4)\left(8314 \frac{\text{J}}{\text{kmol}\cdot\text{K}}\right)(20°\text{C} + 273°)}{29 \frac{\text{kg}}{\text{kmol}}}}$$

$$= 342.9 \text{ m/s}$$

The Mach number is

$$M = \frac{v}{c} = \frac{\left(1700 \frac{\text{km}}{\text{h}}\right)\left(1000 \frac{\text{m}}{\text{km}}\right)}{\left(342.9 \frac{\text{m}}{\text{s}}\right)\left(3600 \frac{\text{s}}{\text{h}}\right)} = 1.38 \quad (1.4)$$

The answer is (B).

2. The speed of sound at the first point in the nozzle is

$$c_1 = \sqrt{kRT}$$

$$= \sqrt{(1.4)\left(287 \frac{\text{J}}{\text{kg}\cdot\text{K}}\right)(40°\text{C} + 273°)}$$

$$= 354.6 \text{ m/s}$$

Calculate the total (stagnation) temperature.

$$\frac{T_0}{T_1} = 1 + \left(\frac{k-1}{2}\right)\text{Ma}_1^2 = 1 + \left(\frac{k-1}{2}\right)\left(\frac{v_1}{c_1}\right)^2$$

$$= 1 + \left(\frac{1.4-1}{2}\right)\left(\frac{300 \frac{\text{m}}{\text{s}}}{354.6 \frac{\text{m}}{\text{s}}}\right)^2$$

$$= 1.143$$

$$T_0 = (1.143)(40°\text{C} + 273°) = 357.8\text{K}$$

The total temperature does not change as the air passes though the nozzle. The air temperature when the Mach number is 2 is

$$\frac{T_0}{T_2} = 1 + \left(\frac{k-1}{2}\right)\text{Ma}_2^2 = 1 + \left(\frac{1.4-1}{2}\right)(2)^2 = 1.8$$

$$T_2 = \frac{357.8\text{K}}{1.8} = 198.8\text{K} \quad (-74°\text{C})$$

The answer is (A).

3. The specific volume of the air is

$$v = \frac{RT}{p}$$

$$= \frac{\left(287 \frac{\text{J}}{\text{kg}\cdot\text{K}}\right)(440\text{K})}{135 \text{ Pa}}$$

$$= 935 \text{ m}^3/\text{kg} \quad (940 \text{ m}^3/\text{kg})$$

The answer is (C).

4. The velocity of air is

$$v = (\text{Ma})c = \text{Ma}\sqrt{kRT}$$

$$= 1.5\sqrt{(1.4)\left(287 \frac{\text{J}}{\text{kg}\cdot\text{K}}\right)(320\text{K})}$$

$$= 537.9 \text{ m/s} \quad (540 \text{ m/s})$$

The answer is (D).

5. The acoustic velocity is

$$c = \sqrt{kRT}$$

For a particular gas, the ratio of specific heats, k, and the specific gas constant, R, are constant. Wall friction will reduce the gas velocity, but it won't affect the acoustic velocity. Only temperature, T, can vary independently.

The answer is (D).

6. The Mach number behind the shockwave is

$$\text{Ma}_2 = \sqrt{\frac{(k-1)\text{Ma}_1^2 + 2}{2k\text{Ma}_1^2 - (k-1)}}$$

$$= \sqrt{\frac{(1.4-1)(1.5)^2 + 2}{(2)(1.4)(1.5)^2 - (1.4-1)}}$$

$$= 0.701 \quad (0.7)$$

The answer is (B).

12 Fluid Machines

PRACTICE PROBLEMS

1. A blower-type fan delivers air at the rate of 110 m³/min against a static gage pressure of 6 cm of water. The air density is 1.2 kg/m³, and the head increase delivered by the pump is 50 m of air.

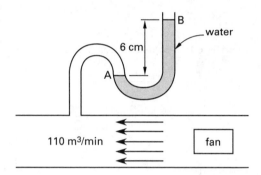

Neglecting velocity pressure, the net power delivered by the fan is most nearly

(A) 0.0016 W
(B) 110 W
(C) 1100 W
(D) 110 000 W

2. A turbine receives steam at a rate of 12 kg/s. At the inlet, the steam is at 200°C and 0.10 MPa (enthalpy of 2875.5 kJ/kg). At the outlet, the steam is at 250°C and 0.20 MPa (enthalpy of 2971.3 kJ/kg). In addition, the steam experiences an increase in kinetic energy of 192 kJ/kg. Assume the turbine is adiabatic. The fluid power imparted by the turbine is most nearly

(A) 200 kW
(B) 350 kW
(C) 1200 kW
(D) 3500 kW

3. A pump that draws 500 kW of power operates at 2000 rpm. The diameter of the pump impeller is 310 mm. A homologous second pump has an impeller diameter of 265 mm and operates at 1400 rpm. Both pumps run water, and the density of the water is constant. The power drawn by the second pump is most nearly

(A) 78 kW
(B) 92 kW
(C) 220 kW
(D) 500 kW

4. A compressor receives superheated steam. The steam is initially at 200°C and 0.20 MPa (entropy of 7.5066 kJ/kg·K). The compressor increases the pressure and temperature of the steam to 0.60 MPa and 370°C. (Steam at 0.60 MPa has an entropy of 7.3724 kJ/kg·K at 300°C, an entropy of 7.5464 kJ/kg·K at 350°C, and an entropy of 7.7079 kJ/kg·K at 400°C.) The thermal efficiency of the compressor is most nearly

(A) 74%
(B) 82%
(C) 85%
(D) 91%

5. Steam is compressed isothermally at a rate of 15 kg/s. The steam is compressed from 0.2 MPa to 0.8 MPa. The inlet temperature of the steam is 250°C, and the efficiency of the compressor is 80%. The molar mass of water is 18.02 kg/kmol. The power driving the compressor is most nearly

(A) 320 kW
(B) 630 kW
(C) 3.2 MW
(D) 6.3 MW

SOLUTIONS

1. The net power delivered is

$$\dot{W}_{\text{fluid}} = \rho g H Q$$

$$= \frac{\left(1.2 \; \frac{\text{kg}}{\text{m}^3}\right)\left(9.81 \; \frac{\text{m}}{\text{s}^2}\right)(50 \; \text{m})\left(110 \; \frac{\text{m}^3}{\text{min}}\right)}{60 \; \frac{\text{s}}{\text{min}}}$$

$$= 1079 \; \text{W} \quad (1100 \; \text{W})$$

The answer is (C).

2. Determine the fluid power imparted by the turbine.

$$\dot{W}_{\text{turb}} = \dot{m}\left(h_e - h_i + \frac{v_e^2 - v_i^2}{2}\right) = \dot{m}(h_e - h_i + \Delta KE)$$

$$= \left(12 \; \frac{\text{kg}}{\text{s}}\right)\left(2971.3 \; \frac{\text{kJ}}{\text{kg}} - 2875.5 \; \frac{\text{kJ}}{\text{kg}} + 192 \; \frac{\text{kJ}}{\text{kg}}\right)$$

$$= 3453.6 \; \text{kJ/s} \quad (3500 \; \text{kW})$$

The answer is (D).

3. Use the similarity law and solve for the power in the second pump. The terms for density are equal and cancel out.

$$\left(\frac{\dot{W}}{\rho N^3 D^5}\right)_2 = \left(\frac{\dot{W}}{\rho N^3 D^5}\right)_1$$

$$\dot{W}_2 = \frac{\dot{W}_1 N_2^3 D_2^5}{N_1^3 D_1^5}$$

$$= \frac{(500 \; \text{kW})\left(1400 \; \frac{\text{rev}}{\text{min}}\right)^3 (265 \; \text{mm})^5}{\left(2000 \; \frac{\text{rev}}{\text{min}}\right)^3 (310 \; \text{mm})^5}$$

$$= 78.3 \; \text{kW} \quad (78 \; \text{kW})$$

The answer is (A).

4. Find the isentropic temperature, T_{es}. This is the temperature at which the pressure is equal to the exit pressure of 0.60 MPa, while the entropy, s, is equal to the input entropy, $s_i = 7.5066$ kJ/kg·K. Interpolate between the entropy values at 300°C and 350°C to find the temperature corresponding to an entropy of 7.5066 kJ/kg·K.

$$T_{es} = T_1 + \left(\frac{s - s_1}{s_2 - s_1}\right)(T_2 - T_1)$$

$$= 300°\text{C} + \left(\frac{7.5066 \; \frac{\text{kJ}}{\text{kg·K}} - 7.3724 \; \frac{\text{kJ}}{\text{kg·K}}}{7.5464 \; \frac{\text{kJ}}{\text{kg·K}} - 7.3724 \; \frac{\text{kJ}}{\text{kg·K}}}\right)$$

$$\times (350°\text{C} - 300°\text{C})$$

$$= 338.56°\text{C}$$

Determine the efficiency of the compressor.

$$\eta_C = \frac{T_{es} - T_i}{T_e - T_i}$$

$$= \frac{338.56°\text{C} - 200°\text{C}}{370°\text{C} - 200°\text{C}}$$

$$= 0.8151 \quad (82\%)$$

The answer is (B).

5. The initial absolute temperature is

$$T_i = 250°\text{C} + 273° = 523\text{K}$$

Use the equation for rate of work for isothermal compression.

$$\dot{W}_{\text{comp}} = \frac{\overline{R} T_i}{M \eta_c} \ln \frac{p_e}{p_i} \dot{m}$$

$$= \frac{\left(8314 \; \frac{\text{J}}{\text{kmol·K}}\right)(523\text{K})}{\left(18.02 \; \frac{\text{kg}}{\text{kmol}}\right)(0.80)} \ln\left(\frac{0.8 \; \text{MPa}}{0.2 \; \text{MPa}}\right)\left(15 \; \frac{\text{kg}}{\text{s}}\right)$$

$$= 6.27 \times 10^6 \; \text{J/s} \quad (6.3 \; \text{MW})$$

The answer is (D).

13 Properties of Substances

PRACTICE PROBLEMS

1. When the volume of an ideal gas is doubled while the temperature is halved, the pressure is

(A) doubled
(B) halved
(C) quartered
(D) quadrupled

2. 1.004 g of superheated ammonia occupy 500 mL. The internal energy of the ammonia is 1.57 kJ. The pressure is four times standard atmospheric pressure. What is most nearly the specific enthalpy of the ammonia?

(A) 1600 kJ/kg
(B) 1800 kJ/kg
(C) 2000 kJ/kg
(D) 2700 kJ/kg

3. 0.8 kg of helium is in a cube-shaped vessel with edges measuring 1.3 m. The pressure is two times standard atmospheric pressure. What is most nearly the temperature of the helium?

(A) 80K
(B) 130K
(C) 160K
(D) 270K

4. A 5.4 kg mixture of nitrogen, oxygen, and butane in a 3 m^3 container is at atmospheric pressure and has a temperature of 300K. What is most nearly the molecular weight of the mixture?

(A) 28 kg/kmol
(B) 32 kg/kmol
(C) 44 kg/kmol
(D) 58 kg/kmol

5. An ideal gas inside a closed system is initially at atmospheric pressure and a temperature of 300K. The pressure is increased by 0.5 bar. Most nearly, what is the final temperature of the gas?

(A) 150K
(B) 200K
(C) 300K
(D) 450K

6. A 1 m long cylindrical vessel with a diameter of 0.5 m contains water in a saturated thermodynamic state. The total internal energy of the water is 36.93 MJ, and the density is approximately 1 g/cm^3. What is most nearly the specific internal energy?

(A) 47 kJ/kg
(B) 190 kJ/kg
(C) 240 kJ/kg
(D) 470 kJ/kg

SOLUTIONS

1. Using the equation of state for an ideal gas,

$$\frac{p_1 v_1}{T_1} = \frac{p_2 v_2}{T_2}$$

$$v_2 = 2v_1$$

$$T_2 = \frac{T_1}{2}$$

$$p_2 = \frac{p_1 v_1 T_2}{T_1 v_2} = \frac{p_1 v_1 \left(\frac{T_1}{2}\right)}{T_1 (2v_1)}$$

$$= \frac{p_1}{4}$$

The pressure is quartered.

The answer is (C).

2. Find the specific internal energy of the ammonia.

$$u = \frac{U}{m}$$

$$= \left(\frac{1.57 \text{ kJ}}{1.004 \text{ g}}\right)\left(1000 \ \frac{\text{g}}{\text{kg}}\right)$$

$$= 1563.7 \text{ kJ/kg}$$

Find the specific volume of the ammonia.

$$v = \frac{V}{m}$$

$$= \frac{(500 \text{ mL})\left(1000 \ \frac{\text{g}}{\text{kg}}\right)}{(1.004 \text{ g})\left(1000 \ \frac{\text{L}}{\text{m}^3}\right)\left(1000 \ \frac{\text{mL}}{\text{L}}\right)}$$

$$= 0.498 \text{ m}^3/\text{kg}$$

The specific enthalpy of the ammonia is

$$h = u + pv$$

$$= 1563.7 \ \frac{\text{kJ}}{\text{kg}} + (4)(101.3 \text{ kPa})\left(0.498 \ \frac{\text{m}^3}{\text{kg}}\right)$$

$$= 1765.5 \text{ kJ/kg} \quad (1800 \text{ kJ/kg})$$

The answer is (B).

3. Atmospheric pressure is 101.3 kPa. From the ideal gas equation, the temperature of the helium is

$$pV = mRT$$

$$T = \frac{pV}{mR}$$

$$= \frac{(2)(101.3 \text{ kPa})(1.3 \text{ m})^3}{(0.8 \text{ kg})\left(\frac{8.314 \ \frac{\text{kJ}}{\text{kmol} \cdot \text{K}}}{4.0026 \ \frac{\text{kg}}{\text{kmol}}}\right)}$$

$$= 267.9 \text{K} \quad (270 \text{K})$$

The answer is (D).

4. Substitute the equation for the specific gas constant into the ideal gas equation. Rearrange to find the molecular weight of the mixture.

$$pV = mRT$$

$$= m\left(\frac{\overline{R}}{\text{mol. wt}}\right)T$$

$$\text{mol. wt} = \frac{m\overline{R}T}{pV}$$

$$= \frac{(5.4 \text{ kg})\left(8.314 \ \frac{\text{kJ}}{\text{kmol} \cdot \text{K}}\right)(300 \text{K})}{(101.3 \text{ kPa})(3 \text{ m}^3)}$$

$$= 44 \text{ kg/kmol}$$

The answer is (C).

5. The temperature of the gas after the increase in pressure is

$$\frac{p_1 v_1}{T_1} = \frac{p_2 v_2}{T_2}$$

$$T_2 = \left(\frac{p_2}{p_1}\right) T_1$$

$$= \left(\frac{101.3 \text{ kPa} + (0.5 \text{ bar})\left(100 \ \frac{\text{kPa}}{\text{bar}}\right)}{101.3 \text{ kPa}}\right)(300 \text{K})$$

$$= 448 \text{K} \quad (450 \text{K})$$

The answer is (D).

6. Find the volume of the vessel.

$$V = \pi r^2 L = \pi \left(\frac{d}{2}\right)^2 L$$
$$= \pi \left(\frac{0.5 \text{ m}}{2}\right)^2 (1 \text{ m})$$
$$= 0.196 \text{ m}^3$$

Find the mass of water in the vessel.

$$m = V\rho$$
$$= \frac{(0.196 \text{ m}^3)\left(1 \frac{\text{g}}{\text{cm}^3}\right)\left(10^6 \frac{\text{cm}^3}{\text{m}^3}\right)}{1000 \frac{\text{g}}{\text{kg}}}$$
$$= 196 \text{ kg}$$

Find the specific internal energy of the water.

$$u = \frac{U}{m}$$
$$= \left(\frac{36.93 \text{ MJ}}{196 \text{ kg}}\right)\left(1000 \frac{\text{kJ}}{\text{MJ}}\right)$$
$$= 188.08 \text{ kJ/kg} \quad (190 \text{ kJ/kg})$$

The answer is (B).

14 Laws of Thermodynamics

PRACTICE PROBLEMS

1. The forebay of a dam feeds 900 Mg/s of water to an electrical generating plant. The water has a specific volume of 0.001 m³/kg. It enters the generators at 900 kPa and leaves at 200 kPa. Kinetic and potential energy changes are insignificant. What is most nearly the power generated by the dam?

(A) 180 MW
(B) 630 MW
(C) 810 MW
(D) 900 MW

2. Which of the following principles is the closest interpretation of the first law of thermodynamics for a closed system?

(A) The mass within a closed control volume does not change.
(B) The net energy crossing the system boundary equals the change in energy inside the system.
(C) The change of total energy is equal to the rate of work performed.
(D) All real processes tend toward increased disorder.

3. What is the origin of the energy conservation equation used in flow systems?

(A) Newton's first law of motion
(B) Newton's second law of motion
(C) the first law of thermodynamics
(D) the second law of thermodynamics

4. Steam flows into a turbine at a rate of 10 kg/s, and 10 kW of heat are lost from the turbine.

	inlet conditions	exit conditions
pressure	1.0 MPa	0.1 MPa
temperature	350°C	–
quality	–	100%

Elevation and kinetic energy effects are negligible. What is most nearly the power output from the turbine?

(A) 4000 kW
(B) 4400 kW
(C) 4800 kW
(D) 5000 kW

5. For which type of process is the equation $dQ = TdS$ valid?

(A) irreversible
(B) isothermal
(C) reversible
(D) isobaric

6. Steam is quickly compressed by a piston in a cylinder with a compression ratio of 4:1. Before compression, the temperature of the steam is 400K, and the steam is at atmospheric pressure. After compression, the temperature of the steam is 600K. What is most nearly the work per unit mass done by the piston?

(A) 300 kJ/kg
(B) 315 kJ/kg
(C) 325 kJ/kg
(D) 330 kJ/kg

7. Steam enters an isentropic turbine at an initial pressure of 800 kPa. As the steam leaves the turbine, its pressure is 120 kPa and its quality is 0.94. The mass flow rate is 1.7 kg/s. Kinetic and potential energy changes are insignificant. What is most nearly the power produced by the turbine?

(A) 370 kW
(B) 570 kW
(C) 600 kW
(D) 960 kW

8. Superheated steam at 300°C enters a turbine at 1 MPa. The mixture of saturated liquid water and steam leaving the turbine is at atmospheric pressure and has a quality of 0.87. No heat is lost in the turbine, and the mass flow rate is 2 kg/s. What is most nearly the work output of the turbine?

(A) 500 kW
(B) 700 kW
(C) 1.1 MW
(D) 1.3 MW

9. 7 kg of neon is stored in a rigid tank at three times atmospheric pressure and a temperature of 70°C. 30 kJ of heat is added to the neon. What is most nearly the final temperature of the neon?

(A) 70°C
(B) 73°C
(C) 74°C
(D) 77°C

10. A device expends 130 kJ of energy while pressurizing 10 kg of water initially at 17°C. The isentropic efficiency of the device is 50%. Inefficiencies are represented by a heat loss from the device casing. What is most nearly the final temperature of the water?

(A) 18°C
(B) 19°C
(C) 20°C
(D) 21°C

11. An adiabatic pump receives 1.5 kg/s of 15 kPa water and discharges it at 15 MPa. The specific volume of the entering water is 0.001055 m³/kg. Consider the water to be incompressible. The isentropic efficiency of the pump is 0.82. Velocity and elevation changes are insignificant. The water does not increase in temperature significantly. Most nearly, what is the minimum electrical power required to drive the pump?

(A) 13 kW
(B) 20 kW
(C) 23 kW
(D) 30 kW

12. An engine operates at a constant temperature of 90°C. Through a reversible process, the engine's work output is 5.3 kJ, and the heat loss is 4.7 kJ. What is most nearly the change in entropy during the process?

(A) 0.013 kJ/K
(B) 0.014 kJ/K
(C) 0.015 kJ/K
(D) 0.016 kJ/K

13. In a heat treating process, a 2 kg metal part (specific heat = 0.5 kJ/kg·K) initially at 800°C is quenched in a tank containing 200 kg of water initially at 20°C. What is most nearly the total entropy change of the process immediately after quenching?

(A) 2.3 kJ/K (decrease)
(B) 0.65 kJ/K (decrease)
(C) 0.90 kJ/K (increase)
(D) 1.4 kJ/K (increase)

SOLUTIONS

1. The generators are driven by hydraulic turbines. Since the process is adiabatic, and since velocity and elevation changes are insignificant, all of the terms in the steady flow energy equation drop out except work and enthalpy. The turbine work per unit mass is

$$w_{\text{turbine}} = h_i - h_e$$

In the absence of compressed water tables (giving enthalpies of subcooled and compressed water), the turbine work must be calculated from more basic principles. Enthalpy is defined as $h = pv + u$, so the turbine work per unit mass is

$$w_{\text{pump}} = h_i - h_e = p_i v_i - p_e v_e$$

Since the water is incompressible, the specific volume is unchanged. The turbine work per unit mass is

$$w_{\text{pump}} = v_i(p_i - p_e)$$

The power generated by turbines is

$$\dot{W} = \dot{m} v_i (p_i - p_e)$$
$$= \left(900 \ \frac{\text{Mg}}{\text{s}}\right)\left(0.001 \ \frac{\text{m}^3}{\text{kg}}\right)(900 \ \text{kPa} - 200 \ \text{kPa})$$
$$= 630 \ \text{MW}$$

The answer is (B).

2. The first law is

$$Q - W = \Delta U$$

Q is the heat transfer into the system across the system boundary. W is the energy transferred across the system boundary to the surroundings in the form of work. ΔU is the change in energy stored within the system in the form of internal energy. This expression shows that the energy transferred across the system boundary comes from a change in stored energy.

The answer is (B).

3. The energy equation is based on the first law of thermodynamics.

The answer is (C).

4. From the steam tables,

$$h_i = 3157.7 \text{ kJ/kg}$$
$$h_e = 2675.5 \text{ kJ/kg}$$
$$\dot{W} = \dot{m}(h_i - h_e) + Q$$
$$= \left(10 \ \frac{\text{kg}}{\text{s}}\right)\left(3157.7 \ \frac{\text{kJ}}{\text{kg}} - 2675.5 \ \frac{\text{kJ}}{\text{kg}}\right) - 10 \text{ kW}$$
$$= 4812 \text{ kW} \quad (4800 \text{ kW})$$

The answer is (C).

5. Since the two sides are separated by an equal sign, the process must be reversible. For an irreversible process, $dQ < T\,ds$.

The answer is (C).

6. The steam is contained in the cylinder, so the process is closed. Since the steam is compressed quickly, the time available for heat loss is low, and this process can be assumed to be adiabatic. Use the first law of thermodynamics for a closed system.

$$W_{\text{out}} = Q - \Delta U = 0 - (U_2 - U_1)$$
$$= U_1 - U_2$$
$$W_{\text{in}} = -W_{\text{out}}$$
$$= U_2 - U_1$$

The steam is initially at $400\text{K} - 273° = 127°\text{C}$. From the saturated steam table, the saturation pressure corresponding to $127°\text{C}$ is higher than 1 atm (0.10135 MPa), so the steam is superheated. (Alternatively, the temperature of $127°\text{C}$ is higher than the saturation temperature corresponding to 1 atm, which is $100°\text{C}$, so the steam is superheated.)

The initial temperature and pressure are $127°\text{C}$ and 0.10135 MPa. For convenience, calculate the internal energy and specific volume of the steam from the superheated water table at 0.10 MPa with an interpolated temperature of $127°\text{C}$.

$$u_1 = 2548 \text{ kJ/kg}$$
$$v_1 = 1.826 \text{ m}^3/\text{kg}$$

For reciprocating machines, the compression ratio is a ratio of volumes, not of pressures. The final specific volume is

$$v_2 = \frac{v_1}{r_v} = \frac{1.826 \ \frac{\text{m}^3}{\text{kg}}}{4} = 0.4565 \text{ m}^3/\text{kg}$$

The final pressure temperature is $600\text{K} - 273° = 327°\text{C}$.

Search the superheated steam table for a combination of these values. The values correspond roughly to a pressure of 0.60 MPa. At 0.60 MPa and $327°\text{C}$, the interpolated internal energy is

$$u_2 \approx 2844 \text{ kJ/kg}$$

The work required to compress the steam is

$$w_{\text{in}} = u_2 - u_1$$
$$= 2844 \ \frac{\text{kJ}}{\text{kg}} - 2548 \ \frac{\text{kJ}}{\text{kg}}$$
$$= 296 \text{ kJ/kg} \quad (300 \text{ kJ/kg})$$

The answer is (A).

7. At the turbine exit, the steam is a liquid-vapor mixture at 120 kPa. From the saturated water table, this roughly corresponds to a saturation temperature of $105°\text{C}$, which will be used for the exit properties. The exit entropy is

$$s = xs_{fg} + s_f$$
$$= (0.94)\left(5.9328 \ \frac{\text{kJ}}{\text{kg·K}}\right) + 1.3630 \ \frac{\text{kJ}}{\text{kg·K}}$$
$$= 6.940 \text{ kJ/kg·K}$$

The enthalpy of the steam at the exit is

$$h_e = xh_{fg} + h_f$$
$$= (0.94)\left(2243.7 \ \frac{\text{kJ}}{\text{kg}}\right) + 440.15 \ \frac{\text{kJ}}{\text{kg}}$$
$$= 2549.23 \text{ kJ/kg}$$

Since the turbine is isentropic, the entrance entropy is the same as the exit entropy, so the entrance state is defined by the pressure (800 kPa) and entropy (6.940 kJ/kg·K). Locate the known entropy in the 0.8 MPa superheated water table. By interpolation, the entrance temperature is approximately $228°\text{C}$, and the entrance enthalpy is approximately 2901 kJ/kg.

Since the turbine is isentropic, it is adiabatic, and the heat loss term is zero. The work performed by the turbine is

$$\dot{W}_{\text{out}} = \dot{m}(h_i - h_e)$$
$$= \left(1.7 \ \frac{\text{kg}}{\text{s}}\right)\left(2901 \ \frac{\text{kJ}}{\text{kg}} - 2549.23 \ \frac{\text{kJ}}{\text{kg}}\right)$$
$$= 598.0 \text{ kW} \quad (600 \text{ kW})$$

The answer is (C).

8. From the superheated steam table at 1 MPa and 300°C, the properties of the steam entering the turbine are

$$h_i = 3051.2 \text{ kJ/kg}$$
$$s_i = 7.1229 \text{ kJ/kg·K}$$

From the saturated water table at 0.10135 kPa (atmospheric), the heat of vaporization, h_{fg}, is 2257.0 kJ/kg. The enthalpy of the liquid-steam mixture leaving the turbine is

$$h_e = h_f + xh_{fg}$$
$$= 419.04 \text{ } \frac{\text{kJ}}{\text{kg}} + (0.87)\left(2257.0 \text{ } \frac{\text{kJ}}{\text{kg}}\right)$$
$$= 2382.6 \text{ kJ/kg}$$

The power generated is

$$\dot{W}_{out} = \dot{m}(h_i - h_e)$$
$$= \frac{\left(2 \text{ } \frac{\text{kg}}{\text{s}}\right)\left(3051.2 \text{ } \frac{\text{kJ}}{\text{kg}} - 2382.6 \text{ } \frac{\text{kJ}}{\text{kg}}\right)}{1000 \text{ } \frac{\text{kW}}{\text{MW}}}$$
$$= 1.337 \text{ MW} \quad (1.3 \text{ MW})$$

The answer is (D).

9. Use the first law of thermodynamics and the equation for change in internal energy to calculate the final temperature. The heat capacity of neon at a constant volume is 0.618 kJ/kg·K.

$$Q - W = \Delta U = mc_v \Delta T = mc_v(T_2 - T_1)$$
$$T_2 = \frac{Q - W}{mc_v} + T_1$$
$$= \frac{30 \text{ kJ} - 0 \text{ kJ}}{(7 \text{ kg})\left(0.618 \text{ } \frac{\text{kJ}}{\text{kg·K}}\right)} + 70°\text{C}$$
$$= 76.93°\text{C} \quad (77°\text{C})$$

The answer is (D).

10. The heat loss and net work are

$$Q_{out} = 0.50 W_{in} = (0.50)(130 \text{ kJ})$$
$$= 65 \text{ kJ}$$

From the first law,

$$Q_{in} - W_{out} = \Delta U$$

Alternatively,

$$W_{in} - Q_{out} = \Delta U$$

For any substance in any state, $dq = c_v \, dT$. The specific heats, c_v and c_p, for liquids are essentially identical. The specific heat is essentially constant over a relatively large range of temperatures. The specific heat of water is 4.18 kJ/kg·K.

$$W_{in} - Q_{out} = m\Delta u = mc_v(T_2 - T_1)$$
$$130 \text{ kJ} - 65 \text{ kJ} = (10 \text{ kg})\left(4.18 \text{ } \frac{\text{kJ}}{\text{kg·K}}\right)(T_2 - 17°\text{C})$$
$$T_2 = 18.56°\text{C} \quad (19°\text{C})$$

The answer is (B).

11. Since the pump is adiabatic, and since velocity and elevation changes are insignificant, all of the terms in the steady flow energy equation drop out except work and enthalpy. The pump work per unit mass is

$$w_{pump} = h_i - h_e$$

In the absence of compressed water tables (giving enthalpies of subcooled and compressed water), the pump work must be calculated from more basic principles. Enthalpy is defined as $h = pv + u$, so the pump work per unit mass is

$$w_{pump} = h_i - h_e = p_i v_i - p_e v_e$$

Since the water is incompressible, the specific volume is unchanged. The pump work per unit mass is

$$w_{pump} = v_i(p_i - p_e)$$

The power required to operate the pump is

$$\dot{W} = \frac{\dot{m} v_i (p_i - p_e)}{\eta_s}$$
$$= \frac{\left(1.5 \text{ } \frac{\text{kg}}{\text{s}}\right)\left(0.001055 \text{ } \frac{\text{m}^3}{\text{kg}}\right)}{0.82}$$
$$\quad \times \left(15 \text{ kPa} - (15 \text{ MPa})\left(10^3 \text{ } \frac{\text{kPa}}{\text{MPa}}\right)\right)$$
$$= -28.9 \text{ kW} \quad (30 \text{ kW})$$

By the standard thermodynamic convention, a negative "work out" represents "work in" (i.e., work being done on the system).

The answer is (D).

12. The change in entropy is

$$\Delta S_{\text{out}} = \frac{Q_{\text{out}}}{T_{\text{reservoir}}}$$
$$= \frac{4.7 \text{ kJ}}{90°\text{C} + 273°}$$
$$= 0.013 \text{ kJ/K}$$

The answer is (A).

13. Calculate the final temperature, T_f.

$$(mc\Delta T)_{\text{metal}} + (mc\Delta T)_{\text{water}} = 0$$
$$(2 \text{ kg})\left(0.5 \ \frac{\text{kJ}}{\text{kg·K}}\right)(800°\text{C} - T_f)$$
$$+ (200 \text{ kg})\left(4.18 \ \frac{\text{kJ}}{\text{kg·K}}\right)(20°\text{C} - T_f)$$
$$= 0$$
$$T_f = 20.93°\text{C} \quad (21°\text{C})$$

For a solid or liquid, the total entropy is

$$\Delta S = mc \ln \frac{T_2}{T_1}$$

So, considering a system consisting of the metal part and the water in the quenching tank,

$$\Delta S_{\text{metal}} = mc \ln \frac{T_2}{T_1}$$
$$= (2 \text{ kg})\left(0.5 \ \frac{\text{kJ}}{\text{kg·K}}\right) \ln \frac{21°\text{C} + 273°}{800°\text{C} + 273°}$$
$$= -1.295 \text{ kJ/K}$$

$$\Delta S_{\text{water}} = mc \ln \frac{T_2}{T_1}$$
$$= (200 \text{ kg})\left(4.18 \ \frac{\text{kJ}}{\text{kg·K}}\right) \ln \frac{21°\text{C} + 273°}{20°\text{C} + 273°}$$
$$= 2.655 \text{ kJ/K}$$

$$\Delta S_{\text{total}} = \Delta S_{\text{metal}} + \Delta S_{\text{water}}$$
$$= -1.295 \ \frac{\text{kJ}}{\text{K}} + 2.655 \ \frac{\text{kJ}}{\text{K}}$$
$$= 1.36 \text{ kJ/K} \quad (1.4 \text{ kJ/K})$$

The answer is (D).

15 Power Cycles and Entropy

PRACTICE PROBLEMS

1. Two Carnot engines operate in series such that the heat rejected from one is the heat input to the other. The heat transfer from the high-temperature reservoir is 500 kJ. The overall temperature limits are 1000K and 400K. Both engines produce equal work. What is most nearly the intermediate temperature between the two engines?

(A) 400K
(B) 500K
(C) 700K
(D) 1000K

2. At the start of compression in an air-standard Otto cycle with a compression ratio of 10, air is at 100 kPa and 40°C. A heat addition of 2800 kJ/kg is made. What is most nearly the thermal efficiency?

(A) 52%
(B) 60%
(C) 64%
(D) 67%

3. A Carnot refrigerator receives heat from a cold reservoir at 0°C. The power input is 1750 W per ton of refrigeration. What is most nearly the system's coefficient of performance?

(A) 1.4
(B) 1.6
(C) 1.8
(D) 2.0

4. Which of the following is a proper statement of the second law of thermodynamics?

(A) It is impossible for a heat engine to produce net work in a complete cycle if it exchanges heat only with bodies at a lower temperature.
(B) It is impossible for a system working in a complete cycle to accomplish, as its sole effect, the transfer of heat from a body at a given temperature to a body at a higher temperature.
(C) It is impossible for a system working in a complete cycle to accomplish, as its sole effect, the transfer of heat from a body at a given temperature to a body at a lower temperature.
(D) It is impossible for a heat engine to produce net work in a complete cycle if it exchanges heat only with bodies exhibiting a temperature differential.

5. How does an adiabatic process compare to an isentropic process?

(A) adiabatic: heat transfer $= 0$; isentropic: heat transfer $\neq 0$
(B) adiabatic: heat transfer $\neq 0$; isentropic: heat transfer $= 0$
(C) adiabatic: reversible; isentropic: not reversible
(D) both: heat transfer $= 0$; isentropic: reversible

6. A Carnot cycle refrigerator operates between -11°C and 22°C. What is the coefficient of performance?

(A) 0.25
(B) 1.1
(C) 4.3
(D) 7.9

7. What are the processes in an ideal Otto combustion cycle?

(A) two constant volume processes and two isentropic processes

(B) two constant pressure processes and two isentropic processes

(C) two constant volume processes and two constant temperature processes

(D) two constant pressure processes and two constant temperature processes

8. Which of the following statements is FALSE?

(A) The Carnot cycle is dependent on the source and sink temperatures, not the working fluid.

(B) The thermal efficiency of a power cycle is defined as the ratio of useful work output to the supplied input energy.

(C) The maximum work obtained from a cycle is dependent on the temperature of the local environment.

(D) Maximum work output will be obtained in an irreversible process.

SOLUTIONS

1. Let T_H, T_I, and T_L represent the high, intermediate, and low temperatures, respectively.

For the two cycles,

$$W_1 = W_2$$
$$\eta_{th,1} Q_{H,1} = \eta_{th,2} Q_{H,2}$$

However,

$$Q_{H,2} = Q_{L,1} = (1 - \eta_{th,1}) Q_{H,1}$$

$$\left(\frac{T_H - T_I}{T_H}\right) Q_{H,1} = \left(\frac{T_I - T_L}{T_I}\right)\left(1 - \frac{T_H - T_I}{T_H}\right) Q_{H,1}$$

$$T_I = \frac{T_H + T_L}{2} = \frac{1000\text{K} + 400\text{K}}{2}$$
$$= 700\text{K}$$

The answer is (C).

2. Find the thermal efficiency.

$$\eta = 1 - r^{1-k} = 1 - 10^{1-1.4}$$
$$= 0.602 \quad (60\%)$$

The answer is (B).

3. The coefficient of performance, COP, is the ratio of the rate of heat transfer to power input. In the SI system, one ton of refrigeration corresponds to 3516 W. Therefore,

$$\text{COP} = \frac{Q_L}{W}$$
$$= \frac{(1 \text{ ton})\left(3516 \frac{\text{W}}{\text{ton}}\right)}{1750 \text{ W}}$$
$$= 2.01 \quad (2.0)$$

The answer is (D).

4. Option B is the Clausius statement of the second law.

The answer is (B).

5. An adiabatic process is one in which there is no heat flow. It is not necessarily reversible. An isentropic process has no heat flow and is reversible.

The answer is (D).

6. The coefficient of performance for a Carnot refrigeration cycle is

$$\text{COP}_c = \frac{T_L}{T_H - T_L}$$
$$= \frac{-11°C + 273°}{(22°C + 273°) - (-11°C + 273°)}$$
$$= 7.9$$

(As can be seen, it is not necessary to convert temperatures to absolute temperatures when calculating temperature differences.)

The answer is (D).

7. An Otto combustion cycle consists of the following processes.

    A to B: isentropic compression

    B to C: constant volume heat addition

    C to D: isentropic expansion

    D to A: constant volume heat rejection

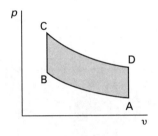 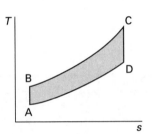

The answer is (A).

8. Maximum work output will be obtained in a reversible process. The difference between the maximum and the actual work output is the process irreversibility.

The answer is (D).

16 Mixtures of Gases, Vapors, and Liquids

PRACTICE PROBLEMS

1. A gas mixture with volumetric proportions of 30% carbon dioxide (specific heat of 0.867 kJ/kg·K) and 70% nitrogen (specific heat of 1.043 kJ/kg·K) is cooled at constant pressure from 150°C to 50°C. What is most nearly the energy released as heat per unit mass of mixture?

(A) 46 kJ/kg

(B) 97 kJ/kg

(C) 160 kJ/kg

(D) 210 kJ/kg

2. By weight, atmospheric air is approximately 23.15% oxygen and 76.85% nitrogen. What is most nearly the partial pressure of oxygen in the air at standard pressure?

(A) 21 kPa

(B) 23 kPa

(C) 26 kPa

(D) 30 kPa

3. In an air-water vapor mixture at 30°C, the partial pressure of the water vapor is 1.5 kPa and the partial pressure of the dry air is 100.5 kPa. The saturation pressure for water vapor at 30°C is 4.246 kPa. The gas constant for water vapor is 0.46152 kJ/kg·K, and the specific gas constant for air is 0.28700 kJ/kg·K. What is most nearly the relative humidity of the air-water vapor mixture?

(A) 0.0090

(B) 0.042

(C) 0.35

(D) 0.65

4. Water at 25°C has a saturation vapor pressure of 3.1504 kPa. Liquid compound B (molecular weight of 52.135) at 25°C has a saturation vapor pressure of 7.2601 kPa. If 75 g of liquid water are mixed with 45 g of liquid compound B, what is most nearly the resulting vapor pressure of the solution?

(A) 3.4 kPa

(B) 3.9 kPa

(C) 4.7 kPa

(D) 5.2 kPa

5. Atmospheric air has a relative humidity of 60% at a dry-bulb temperature of 23°C. What is most nearly the humidity ratio of the air?

(A) 0.011 kg/kg

(B) 0.018 kg/kg

(C) 0.022 kg/kg

(D) 0.300 kg/kg

6. An air-water vapor mixture at 25°C has a partial pressure of water vapor of 1.75 kPa, and a partial pressure of dry air of 100.2 kPa. The humidity ratio of the air-water vapor mixture is 0.01086. The gas constant for water vapor is 0.46152 kJ/kg·K, and the gas constant for dry air is 0.28700 kJ/kg·K. If the mass of water vapor in the mixture is 5 kg, what is most nearly the mass of dry air?

(A) 130 kg

(B) 190 kg

(C) 270 kg

(D) 460 kg

7. Atmospheric air has a wet-bulb temperature of 24°C and a dry-bulb temperature of 31°C at a pressure of 101.325 kPa. What is most nearly the humidity ratio of the sample?

(A) 0.002 kg/kg

(B) 0.006 kg/kg

(C) 0.016 kg/kg

(D) 0.018 kg/kg

8. A closed container is partially filled with a mixture of two volatile liquids. The mixture consists of 2 moles of liquid A and 3 moles of liquid B. The vapor pressures of the pure liquids A and B are 6.0 kPa and 8.5 kPa, respectively. The mixture behaves ideally. At equilibrium, what is the mole fraction of component B in the vapor?

(A) 0.32

(B) 0.59

(C) 0.68

(D) 0.74

9. A gas mixture contains 2 moles of helium, 1 mole of neon, and equal numbers of moles of argon and krypton. The mixture is sealed in a container with a volume of 0.03 m³ and is kept at three times atmospheric pressure at a temperature of 30°C. What is most nearly the mass fraction of argon in the mixture?

(A) 0.08
(B) 0.16
(C) 0.19
(D) 0.39

10. What does Dalton's law of partial pressures state about gases?

(A) The total pressure of a gas mixture is the sum of the individual gases' partial pressures.
(B) The total volume of a nonreactive gas mixture is the sum of the individual gases' volumes.
(C) Each gas of a mixture has the same partial pressure as that of the mixture.
(D) The gas pressure of a mixture is the weighted average of the individual gas pressures.

11. An air-water vapor mixture is at a temperature of 20°C. The partial pressure of water is 2.3 kPa, and the partial pressure of the dry air is 99.1 kPa. The saturation pressure for water vapor at 20°C is 2.339 kPa. What is most nearly the humidity ratio of the air-water vapor mixture?

(A) 0.014
(B) 0.26
(C) 0.35
(D) 0.38

12. Normal air is approximately 78.1% N_2, 20.9% O_2, 0.9% Ar, and 0.1% CO_2 by volume. A closed container with a volume of 1 m³ holds a sample of normal air at atmospheric pressure and a temperature of 20°C. What is most nearly the partial pressure of Ar in the container?

(A) 1.0 kPa
(B) 4.0 kPa
(C) 10 kPa
(D) 21 kPa

13. 1.3 g of helium, 2.4 g of hydrogen, and 3.7 g of oxygen are sealed in a container with a volume of 0.01 m³. The box is heated to 50°C. What is most nearly the pressure exerted on the walls of the container?

(A) 70 kPa
(B) 100 kPa
(C) 320 kPa
(D) 440 kPa

SOLUTIONS

1. The volumetric fractions are given. For ideal gases, the mole and volumetric fractions are the same. However, the gravimetric fractions are needed to calculate the mixture's specific heat. The gravimetric fraction of CO_2 is

$$y_i = \frac{x_i M_i}{\sum x_i M_i}$$

$$y_{CO_2} = \frac{(0.30)\left(44 \ \frac{kg}{kmol}\right)}{(0.30)\left(44 \ \frac{kg}{kmol}\right) + (0.70)\left(28 \ \frac{kg}{kmol}\right)}$$

$$= 0.402$$

The gravimetric fraction of N_2 is

$$y_{N_2} = 1 - y_{CO_2} = 1 - 0.402$$
$$= 0.598$$

From Gibbs rule, the specific heat is gravimetrically weighted.

$$c_p = \sum y_i c_{p,i}$$

$$= (0.402)\left(0.867 \ \frac{kJ}{kg \cdot K}\right)$$

$$+ (0.598)\left(1.043 \ \frac{kJ}{kg \cdot K}\right)$$

$$= 0.972 \ kJ/kg \cdot K$$

The heat released is

$$\frac{Q}{m} = c_p \Delta T = \left(0.972 \ \frac{kJ}{kg \cdot K}\right)(50°C - 150°C)$$

$$= -97.2 \ kJ/kg \quad (97 \ kJ/kg \ released)$$

The answer is (B).

2. Partial pressure is volumetrically weighted, or weighted per mole. Calculate oxygen's mole fraction.

$$x_i = \frac{\dfrac{y_i}{M_i}}{\sum \dfrac{y_i}{M_i}}$$

$$x_{O_2} = \frac{\dfrac{y_{O_2}}{M_{O_2}}}{\dfrac{y_{O_2}}{M_{O_2}} + \dfrac{y_{N_2}}{M_{N_2}}} = \frac{\dfrac{0.2315}{32 \ \frac{kg}{kmol}}}{\dfrac{0.2315}{32 \ \frac{kg}{kmol}} + \dfrac{0.7685}{28 \ \frac{kg}{kmol}}}$$

$$= 0.209$$

Atmospheric pressure is 101.3 kPa. The partial pressure of O_2 is

$$p_{O_2} = x_{O_2} p = 0.209p$$
$$= (0.209)(101.3 \text{ kPa})$$
$$= 21.1 \text{ kPa} \quad (21 \text{ kPa})$$

The answer is (A).

3. The relative humidity is

$$\phi = \frac{p_v}{p_g} = \frac{1.5 \text{ kPa}}{4.246 \text{ kPa}}$$
$$= 0.3533 \quad (0.35)$$

The answer is (C).

4. The number of moles of water is

$$N = \frac{m}{M} = \frac{75 \text{ g}}{18 \frac{\text{g}}{\text{mol}}} = 4.17 \text{ mol}$$

The number of moles of compound B is

$$N = \frac{m}{M} = \frac{45 \text{ g}}{52.135 \frac{\text{g}}{\text{mol}}}$$
$$= 0.863 \text{ mol}$$

The mole fractions are

$$x_{\text{water}} = \frac{4.17 \text{ mol}}{4.17 \text{ mol} + 0.863 \text{ mol}} = 0.828$$
$$x_B = \frac{0.863 \text{ mol}}{4.17 \text{ mol} + 0.863 \text{ mol}} = 0.172$$

The partial pressures are

$$p_{\text{water}} = x_{\text{water}} p^*_{\text{water}} = (0.828)(3.1504 \text{ kPa})$$
$$= 2.6098 \text{ kPa}$$
$$p_B = x_B p^*_B = (0.172)(7.2601 \text{ kPa})$$
$$= 1.2459 \text{ kPa}$$

The total pressure is

$$p = \sum p_i = 2.6098 \text{ kPa} + 1.2459 \text{ kPa}$$
$$= 3.856 \text{ kPa} \quad (3.9 \text{ kPa})$$

The answer is (B).

5. Use the psychrometric chart (SI units). Locate the dry-bulb temperature of 23°C along the bottom of the chart, and find the intersection point with the curve for 60% relative humidity.

The humidity ratio can be found by following the horizontal grid line which intersects this point to the right. The line intersects the vertical humidity ratio scale at 0.0105 kg/kg (0.011 kg/kg).

The answer is (A).

6. The mass of dry air is

$$m_a = \frac{m_v}{\omega} = \frac{5 \text{ kg}}{0.01086}$$
$$= 460.4 \text{ kg} \quad (460 \text{ kg})$$

The answer is (D).

7. Use the psychrometric chart (SI units). Locate the dry-bulb temperature of 31°C on the horizontal axis along the bottom of the chart. Follow the vertical grid line for the dry-bulb temperature up to the point of intersection with the dashed diagonal line for a wet-bulb temperature of 24°C. Follow horizontally to the right to the vertical humidity ratio scale. The humidity ratio of the sample is 0.016 kg/kg.

The answer is (C).

8. The total number of moles of liquid mixture is

$$N = N_A + N_B = 2 \text{ mol} + 3 \text{ mol} = 5 \text{ mol}$$

The mole fraction of component A in the liquid mixture is

$$x_A = \frac{N_A}{N} = \frac{2 \text{ mol}}{5 \text{ mol}} = 0.4$$

The mole fraction of component B in the liquid mixture is

$$x_B = 1 - x_A = 1 - 0.4 = 0.6$$

From Raoult's law, the partial pressures of the vapors are

$$p_A = x_A p^*_A = (0.40)(6.0 \text{ kPa})$$
$$= 2.4 \text{ kPa}$$
$$p_B = x_B p^*_B = (0.60)(8.5 \text{ kPa})$$
$$= 5.1 \text{ kPa}$$

The total pressure above the liquid mixture is

$$p = \sum p_i = p_A + p_B$$
$$= 2.4 \text{ kPa} + 5.1 \text{ kPa}$$
$$= 7.5 \text{ kPa}$$

The mole fraction of component B in the vapor is equal to the partial pressure ratio.

$$x_B = \frac{p_B}{p} = \frac{5.1 \text{ kPa}}{7.5 \text{ kPa}} = 0.68$$

The answer is (C).

9. Use the ideal gas equation to find the total number of moles in the mixture.

$$N_{\text{mixture}} = \frac{PV}{RT}$$

$$= \frac{(3)(101.3 \text{ kPa})(0.03 \text{ m}^3)\left(1000 \frac{\text{mol}}{\text{kmol}}\right)}{\left(8.314 \frac{\text{kJ}}{\text{kmol}\cdot\text{K}}\right)(30°\text{C} + 273°)}$$

$$= 3.6 \text{ mol}$$

The number of moles of krypton is equal to the number of moles of argon.

$$N_{\text{Kr}} = N_{\text{Ar}} = \frac{N_{\text{mixture}} - N_{\text{He}} + N_{\text{Ne}}}{2}$$

$$= \frac{3.6 \text{ mol} - 2 \text{ mol} - 1 \text{ mol}}{2}$$

$$= 0.3 \text{ mol}$$

Find the mole fraction of each of the gases in the mixture.

$$x_{\text{He}} = \frac{N_{\text{He}}}{N_{\text{mixture}}} = \frac{2 \text{ mol}}{3.6 \text{ mol}}$$

$$= 0.553$$

$$x_{\text{Ne}} = \frac{N_{\text{Ne}}}{N_{\text{mixture}}} = \frac{1 \text{ mol}}{3.6 \text{ mol}}$$

$$= 0.276$$

$$x_{\text{Kr}} = x_{\text{Ar}} = \frac{N_{\text{Ar}}}{N_{\text{mixture}}} = \frac{0.3 \text{ mol}}{3.6 \text{ mol}}$$

$$= 0.086$$

Solve for the mass fraction of argon in the mixture.

$$y_{\text{Ar}} = \frac{x_{\text{Ar}} M_{\text{Ar}}}{x_{\text{He}} M_{\text{He}} + x_{\text{Ne}} M_{\text{Ne}} + x_{\text{Ar}} M_{\text{Ar}} + x_{\text{Kr}} M_{\text{Kr}}}$$

$$= \frac{(0.086)\left(40 \frac{\text{g}}{\text{mol}}\right)}{(0.553)\left(4 \frac{\text{g}}{\text{mol}}\right) + (0.276)\left(20 \frac{\text{g}}{\text{mol}}\right)}$$
$$+ (0.086)\left(40 \frac{\text{g}}{\text{mol}}\right) + (0.086)\left(84 \frac{\text{g}}{\text{mol}}\right)$$

$$= 0.187 \quad (0.19)$$

The answer is (C).

10. Dalton's law states that the total pressure of a gas mixture is the sum of the partial pressures.

The answer is (A).

11. The humidity ratio is

$$\omega = 0.622 \frac{p_v}{p_a}$$

$$= (0.622)\left(\frac{2.3 \text{ kPa}}{99.1 \text{ kPa}}\right)$$

$$= 0.014$$

The answer is (A).

12. Use the ideal gas equation to find the total number of moles of air in the container.

$$N_{\text{air}} = \frac{pV}{RT}$$

$$= \frac{(101.3 \text{ kPa})(1 \text{ m}^3)\left(1000 \frac{\text{mol}}{\text{kmol}}\right)}{\left(8.314 \frac{\text{kJ}}{\text{kmol}\cdot\text{K}}\right)(20°\text{C} + 273°)}$$

$$= 41.6 \text{ mol}$$

Find the number of moles of argon in the container.

$$N_{\text{Ar}} = x_{\text{Ar}} N_{\text{air}}$$

$$= (0.009)(41.6 \text{ mol})$$

$$= 0.3743 \text{ mol}$$

Find the mass of argon in the container.

$$m_{\text{Ar}} = N_{\text{Ar}} M_{\text{Ar}}$$

$$= (0.3743 \text{ mol})\left(40 \frac{\text{g}}{\text{mol}}\right)$$

$$= 14.970 \text{ g}$$

Find the partial pressure of argon in the container.

$$p_{\text{Ar}} = \frac{m_{\text{Ar}} R_{\text{Ar}} T}{V}$$

$$= \frac{(14.970 \text{ g})\left(\dfrac{8.314 \frac{\text{kJ}}{\text{kmol}\cdot\text{K}}}{40 \frac{\text{kg}}{\text{kmol}}}\right)(20°\text{C} + 273°)}{(1 \text{ m}^3)\left(1000 \frac{\text{g}}{\text{kg}}\right)}$$

$$= 0.91 \text{ kPa} \quad (1.0 \text{ kPa})$$

The answer is (A).

13. Find the partial pressure of each component.

$$p_i = \frac{m_i R_i T}{V}$$

$$p_{He} = \frac{(1.3 \text{ g})\left(\frac{8.314 \frac{\text{kJ}}{\text{kg·K}}}{4 \frac{\text{kg}}{\text{kmol}}}\right)(50°C + 273°)}{(0.01 \text{ m}^3)\left(1000 \frac{\text{g}}{\text{kg}}\right)}$$

$$= 87.28 \text{ kPa}$$

$$p_{H_2} = \frac{(2.4 \text{ g})\left(\frac{8.314 \frac{\text{kJ}}{\text{kg·K}}}{2 \frac{\text{kg}}{\text{kmol}}}\right)(50°C + 273°)}{(0.01 \text{ m}^3)\left(1000 \frac{\text{g}}{\text{kg}}\right)}$$

$$= 322.25 \text{ kPa}$$

$$p_{O_2} = \frac{(3.7 \text{ g})\left(\frac{8.314 \frac{\text{kJ}}{\text{kg·K}}}{32 \frac{\text{kg}}{\text{kmol}}}\right)(50°C + 273°)}{(0.01 \text{ m}^3)\left(1000 \frac{\text{g}}{\text{kg}}\right)}$$

$$= 31.05 \text{ kPa}$$

The total pressure exerted on the walls of the container is

$$p = \sum p_i$$
$$= p_{He} + p_{H_2} + p_{O_2}$$
$$= 87.28 \text{ kPa} + 322.25 \text{ kPa} + 31.05 \text{ kPa}$$
$$= 440.58 \text{ kPa} \quad (440 \text{ kPa})$$

The answer is (D).

17 Combustion

PRACTICE PROBLEMS

1. The chemical equation for ethane combustion is

$$7O_2 + 2C_2H_6 \rightarrow 6H_2O + 4CO_2$$

The gases behave ideally. Most nearly, what volume of O_2 at 298K and 1.0 atm is required for complete combustion of 10 L of C_2H_6 (gas) at 500K and 1 atm?

(A) 16 L
(B) 19 L
(C) 21 L
(D) 22 L

2. Complete combustion of 13.02 g of a compound (C_xH_y) produces 40.94 g of CO_2 and 16.72 g of H_2O. What is the empirical formula of the compound?

(A) CH
(B) CH_2
(C) CH_4
(D) CH_2O

3. What are the products of complete combustion of a gaseous hydrocarbon?

I. carbon monoxide
II. carbon dioxide
III. water
IV. ammonia

(A) I only
(B) II and III
(C) I, II, and III
(D) I, III, and IV

4. Theoretically, how many kilograms of air are needed to completely burn 5 kg of ethane (C_2H_6) gas?

(A) 0.8 kg
(B) 19 kg
(C) 81 kg
(D) 330 kg

5. Eleven grams of propane (molecular weight of 44 g/mol) are burned with just enough pure oxygen for complete combustion. What is most nearly the mass of combustion products produced?

(A) 31 g
(B) 39 g
(C) 41 g
(D) 51 g

6. Excess air is required in combustion because it

(A) allows the reaction to occur stoichiometrically
(B) reduces air pollution
(C) reduces the heat requirements
(D) allows complete combustion

7. Octane (C_8H_{18}) is burned with 150% theoretical air in a steady combustion process. Combustion is complete. The total pressure is 1 atm. What is most nearly the air-fuel ratio?

(A) 10 kg air/kg fuel
(B) 15 kg air/kg fuel
(C) 23 kg air/kg fuel
(D) 31 kg air/kg fuel

SOLUTIONS

1. Calculate the number of moles of ethane in the reaction.

$$N_{C_2H_6} = \frac{pV}{RT}$$

$$= \frac{(1.0 \text{ atm})(10 \text{ L})}{\left(0.0821 \frac{\text{atm·L}}{\text{mol·K}}\right)(500\text{K})}$$

$$= 0.24 \text{ mol}$$

The chemical equation coefficients represent molecules, volumes, and moles.

$$\frac{N_{O_2}}{N_{C_2H_6}} = 7 \text{ mol}/2 \text{ mol}$$

The number of moles of oxygen in the reaction is

$$N_{O_2} = \left(\frac{7 \text{ mol}}{2 \text{ mol}}\right)(0.24 \text{ mol})$$

$$= 0.85 \text{ mol}$$

From the ideal gas law, the volumes are directly related to the absolute temperatures.

$$\frac{V_{298K}}{V_{STP}} = \frac{T_{298K}}{T_{STP}} = \frac{298K}{273K}$$

$$= 1.09$$

The volume of 1 mole of ideal gas at standard temperature and pressure (STP) is 22.4 L. The volume of O_2 required at 298K is

$$V_{O_2} = (0.85 \text{ mol})(1.09)\left(22.4 \frac{\text{L}}{\text{mol}}\right)$$

$$= 20.8 \text{ L} \quad (21 \text{ L})$$

The answer is (C).

2. Use x to represent the number of carbon atoms in the fuel molecule. For the combustion of carbon, $C + O_2 \rightarrow CO_2$, each atom of carbon uses one molecule of oxygen and produces one molecule of carbon dioxide. Let y represent the number of hydrogen atoms in the fuel molecule. From the combustion of hydrogen, $2H_2 + O_2 \rightarrow 2H_2O$, each atom of hydrogen gas uses one half of an oxygen atom (one quarter of an oxygen molecule) and produces one half of a water molecule. The stoichiometric equation for combustion is

$$C_xH_y + \left(x + \frac{y}{4}\right)O_2 \rightarrow xCO_2 + \frac{y}{2}H_2O$$

$$x = N_{CO_2} = N_C = \frac{m}{M} = \frac{40.94 \text{ g}}{44 \frac{\text{g}}{\text{mol}}}$$

$$= 0.93 \text{ mol}$$

$$\frac{y}{2} = N_{H_2O} = \frac{16.72 \text{ g}}{18 \frac{\text{g}}{\text{mol}}} = 0.93 \text{ mol}$$

$$= N_H/2$$

$$y = N_H = 1.86 \text{ mol}$$

$$\frac{x}{y} = \frac{N_C}{N_H} = \frac{0.93 \text{ mol}}{1.86 \text{ mol}}$$

$$= 1/2$$

The empirical formula for the compound is CH_2.

The answer is (B).

3. A gaseous hydrocarbon reacts with oxygen to form carbon dioxide and water. Carbon monoxide only forms with incomplete combustion.

The answer is (B).

4. The balanced reaction equation is

$$2C_2H_6 + 7O_2 \rightarrow 4CO_2 + 6H_2O$$

combining weights: $(2)(30) \quad (7)(32) \quad (4)(44) \quad (6)(18)$

The ratio of oxygen to ethane combining masses is

$$\frac{O_2}{C_2H_6} = \frac{(7)(32)}{(2)(30)} = \frac{x}{5 \text{ kg}}$$

$$x = 18.67 \text{ kg}$$

Air is 23.15% O_2 by weight. The mass of air required is

$$\frac{18.67 \text{ kg}}{0.2315} = 80.6 \text{ kg} \quad (81 \text{ kg})$$

The answer is (C).

5. The balanced combustion reaction equation is

$$C_3H_8 + 5O_2 \rightarrow 3CO_2 + 4H_2O$$

combining weights: $44 \quad (5)(32) \quad (3)(44) \quad (4)(18)$

The mass of combustion products produced when 44 g of propane is burned is

$$m_{products} = (3)(44 \text{ g}) + (4)(18 \text{ g}) = 204 \text{ g}$$

The mass of combustion products produced when 11 g of propane is burned can be found by simple proportion.

$$\frac{204 \text{ g}}{44 \text{ g}} = \frac{x}{11 \text{ g}}$$

$$x = 51 \text{ g}$$

The answer is (D).

6. Excess air is required for complete combustion to occur.

The answer is (D).

7. The stoichiometric combustion reaction equation is

$$C_8H_{18} + 12.5(O_2 + 3.76N_2)$$
$$\rightarrow 8CO_2 + 9H_2O + (12.5)(3.76)N_2$$

With 150% theoretical air,

$$C_8H_{18} + (1.5)(12.5)(O_2 + 3.76)N_2$$
$$\rightarrow 8CO_2 + 9H_2O + (1.5)(12.5)(3.76)N_2$$
$$+ (0.5)(12.5)O_2$$

Alternatively,

$$C_8H_{18} + 18.75O_2 + 70.5N_2$$
$$\rightarrow 8CO_2 + 9H_2O + 70.5N_2 + 6.25O_2$$

The coefficients represent the number of molecules, volumes, and moles. The air-fuel ratio is

$$A/F = \frac{\text{mass of air}}{\text{mass of fuel}}$$

$$= \frac{(18.75 \text{ kmol})\left(32 \ \frac{\text{kg}}{\text{kmol}}\right) + (70.5 \text{ kmol})\left(28.014 \ \frac{\text{kg}}{\text{kmol}}\right)}{(8 \text{ kmol})\left(12.01 \ \frac{\text{kg}}{\text{kmol}}\right) + (9 \text{ kmol})\left(2.016 \ \frac{\text{kg}}{\text{kmol}}\right)}$$

$$= 22.54 \text{ kg air/kg fuel} \quad (23 \text{ kg air/kg fuel})$$

The answer is (C).

18 Heating, Ventilating, and Air Conditioning (HVAC)

PRACTICE PROBLEMS

1. A room's sensible space load is 72.5 kW. The latent load from occupants and infiltration, but excluding intentional ventilation, is 3.2 kW. A total of 600 L/s of outside air is required. What is most nearly the room sensible heat ratio?

(A) 0.12
(B) 0.39
(C) 0.96
(D) 1.1

2. The inside design conditions for a conditioned space are 20.1°C dry-bulb and 15.2°C wet-bulb. The air temperature increases 10.6°C as it passes through the conditioned space. The apparatus dew point is 11.9°C. The heat transfer rate is 456 kW. What is most nearly the volumetric flow rate of the air passing through the space?

(A) 12 m^3/s
(B) 19 m^3/s
(C) 22 m^3/s
(D) 36 m^3/s

3. The inside design conditions for a conditioned space are 23.9°C dry-bulb and 16.9°C wet-bulb. The dry-bulb temperature as it enters the air-conditioned space is 12°C. The room sensible heat ratio is 0.89. What is most nearly the wet-bulb temperature of the air entering the space?

(A) 9.0°C
(B) 10°C
(C) 11°C
(D) 12°C

4. The inside design conditions for a conditioned space are 19.5°C dry-bulb and 17.8°C wet-bulb. The outside air is at 31.5°C dry-bulb and 26.7°C wet-bulb. The dry-bulb temperature of the air entering the space is 12.5°C. The apparatus dew point is 10.0°C. Air leaves the conditioner saturated. What is most nearly the system bypass ratio?

(A) 0.11
(B) 0.26
(C) 0.35
(D) 0.48

5. A recirculating air bypass system is used to deliver conditioned air to a room at a rate of 5.8 m^3/s. Air leaves the air conditioner saturated at 13.5°C dry-bulb. The indoor design dry-bulb and wet-bulb temperatures are 24.1°C and 18.2°C, respectively. The system bypass factor is 0.34. After the conditioned air is mixed with the bypassed air, what is most nearly the dry-bulb temperature of the air entering the space?

(A) 15°C
(B) 17°C
(C) 18°C
(D) 21°C

6. An auditorium is designed to seat 4500 people. The ventilation rate is 4.54×10^5 m^3/h of outside air. The outside temperature is −18°C dry-bulb, and the outside pressure is 100.6 kPa. Air leaves the auditorium at 21°C dry-bulb. The sensible heat load from the occupants is 297 kW. The specific heat of air is 1.0048 kJ/kg·K. There is no recirculation, and the auditorium's temperature is maintained by varying the amount of outside air. The temperature at which the air should enter the auditorium is most nearly

(A) 11°C
(B) 13°C
(C) 17°C
(D) 19°C

7. An auditorium is designed to seat 200 people. The ventilation rate is 2.50×10^4 m³/h of outside air, or 3.3×10^4 kg/h. The outside temperature is $-10°C$ dry-bulb, and the outside pressure is 101.3 kPa. Air leaves the auditorium at $23°C$ dry-bulb. The air temperature entering the auditorium is $21.6°C$. Most nearly, how much heat should be supplied to the ventilation air?

(A) 120 kW
(B) 290 kW
(C) 540 kW
(D) 780 kW

8. A batch of green lumber is dried by passing air at $50°C$ dry-bulb and $20°C$ wet-bulb over it in an unheated evaporation booth. The lumber decreases in mass at the rate of 2.6 kg/h. The air leaving the dryer is at 80% relative humidity. Most nearly, what is the incoming air volume?

(A) 170 m³/h
(B) 220 m³/h
(C) 270 m³/h
(D) 350 m³/h

9. Atmospheric air leaves an air conditioning coil saturated at $10°C$. The air passes through a reheater, which raises the temperature sensibly to $22°C$. Most nearly, what is the saturation percentage of the air leaving the reheater?

(A) 47%
(B) 61%
(C) 84%
(D) 100%

SOLUTIONS

1. The room sensible heat ratio is

$$\text{SHR} = \frac{\dot{Q}_\text{sensible}}{\dot{Q}_\text{sensible} + \dot{Q}_\text{latent}}$$
$$= \frac{72.5 \text{ kW}}{72.5 \text{ kW} + 3.2 \text{ kW}}$$
$$= 0.96$$

The answer is (C).

2. The dry-bulb temperature of the air as it enters the conditioned space is

$$T_\text{in} = T_i - 10.6°C = 20.1°C - 10.6°C = 9.5°C$$

Calculate the airflow through the space.

$$\dot{V} = \frac{\dot{Q}}{\rho c_p \Delta T}$$
$$= \frac{456 \text{ kW}}{\left(1.2 \frac{\text{kg}}{\text{m}^3}\right)\left(1 \frac{\text{kJ}}{\text{kg·K}}\right)(20.1°C - 9.5°C)}$$
$$= 35.8 \text{ m}^3/\text{s} \quad (36 \text{ m}^3/\text{s})$$

The answer is (D).

3. Locate the room inside design condition ($23.9°C$ dry-bulb, $16.9°C$ wet-bulb) on the psychrometric chart (SI units). Use the sensible heat ratio protractor on the chart to determine the psychrometric chart slope corresponding to a sensible heat ratio of 0.89. Draw a line (the condition line) back (to the left) with a sensible heat ratio of 0.89. The point corresponding to the room entrance conditions is where the condition line crosses the dry-bulb vertical scale of $12°C$. Follow the diagonal web-bulb temperature lines up and to the left to $11.2°C$ ($11°C$).

The answer is (C).

4. The room's indoor design conditions are the properties of the air that is withdrawn from the room. These are the same properties of the air that is recirculated, bypasses the air conditioner, and is mixed with the conditioned air.

Air-mixing problems are usually solved graphically using the psychrometric chart. A straight line is drawn between the points representing the two airflows, and the bypass factor calculated as the ratio of the bypass volume (an approximation of the ratio of mixing masses) to total mixture volume.

That could be done in this problem, but since the dry-bulb temperatures of the mixture and the two mixing airflows are given, it is easier to calculate the bypass factor directly using the lever rule. This can be done because the dry-bulb temperature is a linear scale on the psychrometric chart, and it doesn't matter if the other scales are inclined.

$$\text{BF} = \frac{T_{\text{mixture}} - T_{\text{conditioned}}}{T_{\text{room}} - T_{\text{conditioned}}}$$
$$= \frac{12.5°C - 10.0°C}{19.5°C - 10.0°C}$$
$$= 0.263 \quad (0.26)$$

The answer is (B).

5. The room's indoor design conditions are the properties of the air that is withdrawn from the room. These are the same properties of the air that is recirculated, bypasses the air conditioner, and is mixed with the conditioned air.

Air-mixing problems are usually solved graphically using the psychrometric chart. A straight line is drawn between the points representing the two airflows, and a ratio of the mixing volumes (an approximation of the ratio of mixing masses) to the total mixture volume (i.e., the bypass factor) is used to locate the mixture point.

That could be done in this problem, but since the mixture dry-bulb temperature is needed, and the dry-bulb temperatures of the two mixing airflows are given, it is easier to calculate the dry-bulb temperature directly using the lever rule. This can be done because the dry-bulb temperature is a linear scale on the psychrometric chart, and it doesn't matter if the other scales are inclined.

$$T_{\text{mixture}} = T_{\text{conditioned}} + \text{BF}(T_{\text{room}} - T_{\text{conditioned}})$$
$$= 13.5°C + (0.34)(24.1°C - 13.5°C)$$
$$= 17.1°C \quad (17°C)$$

The answer is (B).

6. The absolute temperature of outside air is

$$T = -18°C + 273° = 255K$$

There is no moisture in the air at $-18°C$. From the ideal gas law, the density of outside air is

$$\rho = \frac{p}{RT} = \frac{(100.6 \text{ kPa})\left(1000 \frac{\text{Pa}}{\text{kPa}}\right)}{\left(287.03 \frac{\text{J}}{\text{kg·K}}\right)(255K)} = 1.374 \text{ kg/m}^3$$

The mass flow rate is

$$\dot{m} = \rho \dot{V} = \left(1.374 \frac{\text{kg}}{\text{m}^3}\right)\left(4.54 \times 10^5 \frac{\text{m}^3}{\text{h}}\right)$$
$$= 6.238 \times 10^5 \text{ kg/h}$$

The air temperature entering the auditorium is

$$\dot{Q} = \dot{m}c_p(T_{\text{out,air}} - T_{\text{in,air}})$$
$$T_{\text{in,air}} = T_{\text{out,air}} - \frac{\dot{Q}}{\dot{m}c_p}$$
$$= 21°C - \frac{(297 \text{ kW})\left(3600 \frac{\text{s}}{\text{h}}\right)}{\left(6.238 \times 10^5 \frac{\text{kg}}{\text{h}}\right)\left(1.0048 \frac{\text{kJ}}{\text{kg·K}}\right)}$$
$$= 19.3°C \quad (19°C)$$

The answer is (D).

7. The specific heat of dry air is approximately 1.0 kJ/kg·K. The heat needed to heat dry ventilation air from $-10°C$ to $21.6°C$ is

$$\dot{Q} = \dot{m}c_p \Delta T$$
$$= \frac{\left(3.3 \times 10^4 \frac{\text{kg}}{\text{h}}\right)\left(1.0 \frac{\text{kJ}}{\text{kg·°C}}\right)(21.6°C - (-10°C))}{3600 \frac{\text{s}}{\text{h}}}$$
$$= 289.7 \text{ kW} \quad (290 \text{ kW})$$

The answer is (B).

8. Use the psychrometric chart (SI units). Locate the starting point at 50°C dry-bulb and 20°C wet-bulb. The humidity ratio is approximately 0.0024 kg/kg dry air. The specific volume is approximately 0.918 m³/kg dry air.

Since no heat is applied, this is an adiabatic (saturation) process. On the psychrometric chart, the air follows a path of constant enthalpy up and to the left. (Lines of constant enthalpy are almost parallel to lines of constant wet-bulb temperature.) Locate the ending point at 80% relative humidity along a line of constant enthalpy. The humidity ratio is approximately 0.0135 kg/kg dry air.

The incoming air mass flow rate is

$$\dot{m}_a = \frac{\dot{m}_w}{\Delta \omega} = \frac{2.6 \frac{\text{kg}}{\text{h}}}{0.0135 \frac{\text{kg}}{\text{kg}} - 0.0024 \frac{\text{kg}}{\text{kg}}}$$
$$= 234.2 \text{ kg/h}$$

The incoming air volume is

$$\dot{V}_a = \frac{\dot{m}}{\rho_a} = \dot{m}_a v_a$$
$$= \left(234.2 \frac{\text{kg}}{\text{h}}\right)\left(0.918 \frac{\text{m}^3}{\text{kg}}\right)$$
$$= 215 \text{ m}^3/\text{h} \quad (220 \text{ m}^3/\text{h})$$

The answer is (B).

9. Saturation percentage is the same as relative humidity. Use the psychrometric chart (SI units). Locate the starting point (saturated at 10°C). At that point, the relative humidity is 100%. Sensible heating is represented by a horizontal straight line on the psychrometric chart. Move to the right to a dry-bulb temperature of 22°C. At that point, the relative humidity is approximately 47%.

The answer is (A).

19 Conduction

PRACTICE PROBLEMS

1. A wall is constructed of 25 mm thick white pine boards and insulated with 75 mm thick fiberglass insulation. The thermal conductivities of the two materials are 0.11 W/m·K and 0.048 W/m·K, respectively. The internal and external wall temperatures are 22°C and −10°C, respectively.

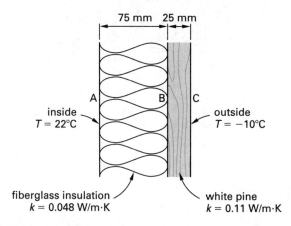

What is most nearly the rate of heat loss per unit area of wall?

(A) 6.7 W/m^2

(B) 18 W/m^2

(C) 19 W/m^2

(D) 35 W/m^2

2. A composite wall constructed of a slab of 2.5 cm of steel ($k = 60.5$ W/m·K) and a slab of 5.0 cm of aluminum ($k = 177$ W/m·K) separates two liquids. The liquid on the steel side has a film coefficient of 15 W/m^2·K and a temperature of 400°C. The liquid on the aluminum side has a film coefficient of 30 W/m^2·K and a temperature of 100°C. Assuming steady-state conditions, what is the approximate temperature at the steel-aluminum interface?

(A) 420K

(B) 470K

(C) 520K

(D) 570K

3. Heat transfer occurs by conduction through a composite wall as shown. The air temperature on one side of the wall is −5°C; on the other, the air temperature is 25°C. The thermal conductivities and film coefficients are 0.065 W/m·K and 0.059 W/m·K, and 30 W/m^2·K and 12 W/m^2·K, as shown.

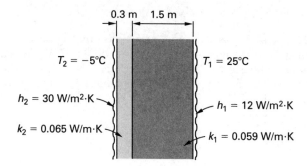

What is most nearly the rate of heat transfer through the wall per unit area?

(A) 1.0 W/m^2

(B) 3.1 W/m^2

(C) 9.8 W/m^2

(D) 17 W/m^2

4. In order to use the lumped capacitance model to evaluate transient heat transfer, the Biot number must be

(A) less than 0.1

(B) less than 1

(C) greater than 0.1

(D) greater than 1

5. A two-layer wall has a total exposed surface area of 120 m^2. The temperature of the hotter side is 550°C; the temperature of the cooler side is 200°C. The layer closer to the hotter side is 0.3 m thick and has a thermal conductivity of 200 W/m·K. The layer closer to the cooler side is 0.5 m thick and has a thermal conductivity of 50 W/m·K.

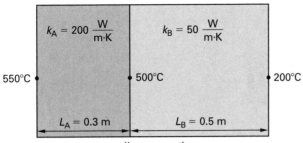

wall cross section

Most nearly, what is the thermal resistance of the wall?

(A) 2.7×10^{-5} K/W

(B) 6.3×10^{-5} K/W

(C) 7.1×10^{-5} K/W

(D) 9.6×10^{-5} K/W

6. The steady-state heat transfer rate through a 10 cm thick by 1 m² homogeneous wall section is 10 W. What is most nearly the temperature difference between the two wall surfaces if the thermal conductivity of the wall material is 1 W/m·°C?

(A) 1.0°C

(B) 2.3°C

(C) 7.8°C

(D) 10°C

7. A 5 cm diameter copper sphere with a uniform temperature of 1000K is suddenly exposed to a fluid with a constant temperature of 300K. The film coefficient is 400 W/m²·K. The density and specific heat of copper are 8933 kg/m³ and 385 J/kg·K, respectively. Using the lumped capacitance method, the temperature of the sphere after 30 seconds is most nearly

(A) 650K

(B) 760K

(C) 850K

(D) 950K

8. Heat flows steadily through a composite wall made up of two materials, A and B, of equal thickness. The thermal conductivity of material A is twice that of material B.

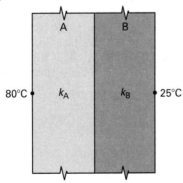

If the temperature of the outside surfaces of A and B are 80°C and 25°C, respectively, the temperature of the interface surface between materials A and B is most nearly

(A) 54°C

(B) 58°C

(C) 62°C

(D) 66°C

9. A 4 cm diameter insulated steel pipe with 3 cm of insulation carries steam at 120°C. The thermal conductivity of the insulation is 0.04 W/m·°C. Neglect the resistance to heat transfer from the steel pipe.

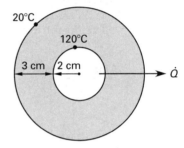

What is most nearly the thermal resistance of the insulation per unit length of pipe?

(A) 0.90 °C/W

(B) 3.6 °C/W

(C) 9.9 °C/W

(D) 11 °C/W

10. A solid copper sphere (8 cm diameter) at 400K is suddenly exposed to air at 25K. The average convective film coefficient is 25 W/m²·K. The density, specific heat, and thermal conductivity of copper are 8933 kg/m³, 410 J/kg·K, and 380 W/m·K, respectively. What is most nearly the time required to cool the sphere to 200K?

(A) 12 min

(B) 16 min

(C) 20 min

(D) 25 min

11. Two long pieces of copper wire (1.6 mm diameter) are connected end to end with a hot soldering iron, forming an infinite cylindrical fin. The thermal conductivity of the copper is 372.1 W/m·K. The minimum melting temperature of the solder is 230K. The surrounding air temperature is 27K. The unit film coefficient is 17 W/m²·K. Disregard radiation losses. The minimum rate of heat application to keep the solder molten is most nearly

(A) 1.3 W

(B) 3.3 W

(C) 6.0 W

(D) 14 W

SOLUTIONS

1. For a composite wall,

$$\dot{Q} = \frac{\Delta T}{R_{\text{total}}}$$

$$= \frac{-(T_C - T_A)}{R_1 + R_2}$$

$$= \frac{-(T_C - T_A)}{\frac{L_1}{k_1 A} + \frac{L_2}{k_2 A}}$$

$$\frac{\dot{Q}}{A} = \frac{-(T_C - T_A)}{\frac{L_1}{k_1} + \frac{L_2}{k_2}}$$

$$= \frac{-(-10°C - 22°C)}{\frac{0.075 \text{ m}}{0.048 \frac{W}{m \cdot K}} + \frac{0.025 \text{ m}}{0.11 \frac{W}{m \cdot K}}}$$

$$= 17.88 \text{ W/m}^2 \quad (18 \text{ W/m}^2)$$

The answer is (B).

2. Determine the total thermal resistance.

$$R = \sum \frac{1}{hA} + \sum \frac{L}{kA}$$

$$RA = \frac{1}{h_{\text{aluminum}}} + \frac{L_{\text{aluminum}}}{k_{\text{aluminum}}} + \frac{L_{\text{steel}}}{k_{\text{steel}}} + \frac{1}{h_{\text{steel}}}$$

$$= \frac{1}{30 \frac{W}{m^2 \cdot K}} + \frac{5.0 \text{ cm}}{\left(177 \frac{W}{m \cdot K}\right)\left(100 \frac{cm}{m}\right)}$$

$$+ \frac{2.5 \text{ cm}}{\left(60.5 \frac{W}{m \cdot K}\right)\left(100 \frac{cm}{m}\right)} + \frac{1}{15 \frac{W}{m^2 \cdot K}}$$

$$= 0.1007 \text{ m}^2 \cdot \text{K/W}$$

The aluminum and steel contribute little to the overall thermal resistance.

Calculate the conductive heat transfer per unit area.

$$\dot{Q} = \frac{\Delta T}{R}$$

$$\frac{\dot{Q}}{A} = \frac{T_{\text{hot}} - T_{\text{cold}}}{RA}$$

$$= \frac{(400°C + 273°) - (100°C + 273°)}{0.1007 \frac{m^2 \cdot K}{W}}$$

$$= 2979 \text{ W/m}^2$$

Calculate the thermal resistance from the steel-side fluid to the interface.

$$RA = \frac{L_{\text{steel}}}{k_{\text{steel}}} + \frac{1}{h_{\text{steel}}}$$

$$= \frac{0.025 \text{ m}}{60.5 \frac{W}{m \cdot K}} + \frac{1}{15 \frac{W}{m^2 \cdot K}}$$

$$= 0.06708 \text{ m}^2 \cdot \text{K/W}$$

The heat transfer is known. Solve the conduction equation for the unknown temperature.

$$\frac{\dot{Q}}{A} = \frac{T_{\text{hot}} - T_{\text{interface}}}{RA}$$

$$2979 \frac{W}{m^2} = \frac{(400°C + 273°) - T_{\text{interface}}}{0.06708 \frac{m^2 \cdot K}{W}}$$

$$T_{\text{interface}} = 473\text{K} \quad (470\text{K})$$

The answer is (B).

3. Determine the total thermal resistance.

$$R = \frac{1}{h_1 A} + \frac{L_1}{k_1 A} + \frac{L_2}{k_2 A} + \frac{1}{h_2 A}$$

$$RA = \frac{1}{h_1} + \frac{L_1}{k_1} + \frac{L_2}{k_2} + \frac{1}{h_2}$$

$$= \frac{1}{12 \frac{W}{m^2 \cdot K}} + \frac{1.5 \text{ m}}{0.059 \frac{W}{m \cdot K}}$$

$$+ \frac{0.3 \text{ m}}{0.065 \frac{W}{m \cdot K}} + \frac{1}{30 \frac{W}{m^2 \cdot K}}$$

$$= 30.16 \text{ m}^2 \cdot \text{K/W}$$

Determine the conductive heat transfer per unit area.

$$\dot{Q} = \frac{T_1 - T_2}{R}$$

$$\frac{\dot{Q}}{A} = \frac{T_1 - T_2}{RA}$$

$$= \frac{25°C - (-5°C)}{30.16 \frac{m^2 \cdot K}{W}}$$

$$= 0.995 \text{ W/m}^2 \quad (1.0 \text{ W/m}^2)$$

The answer is (A).

4. The Biot number is a ratio of the conductive (internal) thermal resistance to the convective (external) thermal resistance.

$$\text{Bi} = \frac{hV}{kA_s}$$

When the Biot number is much less than 1.0 (typically 0.1 or less), surface resistance dominates and internal resistance can be neglected.

The answer is (A).

5. The thermal resistance is

$$R = R_A + R_B = \frac{L_A}{k_A A} + \frac{L_B}{k_B A}$$

$$= \left(\frac{1}{A}\right)\left(\frac{L_A}{k_A} + \frac{L_B}{k_B}\right)$$

$$= \left(\frac{1}{120 \text{ m}^2}\right)\left(\frac{0.3 \text{ m}}{200 \frac{\text{W}}{\text{m}\cdot\text{K}}} + \frac{0.5 \text{ m}}{50 \frac{\text{W}}{\text{m}\cdot\text{K}}}\right)$$

$$= 9.58 \times 10^{-5} \text{ K/W} \quad (9.6 \times 10^{-5} \text{ K/W})$$

The answer is (D).

6. Solve the conduction equation for the temperature difference.

$$\dot{Q} = -kA\frac{dT}{dx} = kA\frac{\Delta T}{\Delta x}$$

$$\Delta T = \frac{\dot{Q}\Delta x}{kA}$$

$$= \frac{(10 \text{ W})(10 \text{ cm})}{\left(1 \frac{\text{W}}{\text{m}\cdot°\text{C}}\right)(1 \text{ m}^2)\left(100 \frac{\text{cm}}{\text{m}}\right)}$$

$$= 1.0°\text{C}$$

The answer is (A).

7. Calculate β, the reciprocal of the time constant.

$$\beta = \frac{hA_s}{\rho V c_p} = \frac{h\pi d^2}{\rho\left(\frac{\pi}{6}\right)d^3 c_p} = \frac{6h}{\rho d c_p}$$

$$= \frac{(6)\left(400 \frac{\text{W}}{\text{m}^2\cdot\text{K}}\right)\left(100 \frac{\text{cm}}{\text{m}}\right)}{\left(8933 \frac{\text{kg}}{\text{m}^3}\right)(5 \text{ cm})\left(385 \frac{\text{J}}{\text{kg}\cdot\text{K}}\right)}$$

$$= 0.01396 \text{ 1/s}$$

The temperature variation of the body with time is

$$T - T_\infty = (T_i - T_\infty)e^{-\beta t}$$

Solve for T.

$$T = T_\infty + (T_i - T_\infty)e^{-\beta t}$$

$$= 300\text{K} + (1000\text{K} - 300\text{K})e^{-(0.01396 \text{ 1/s})(30 \text{ s})}$$

$$= 760.5\text{K} \quad (760\text{K})$$

The answer is (B).

8. For conduction through a plane wall, with T_2 as the intermediate temperature,

$$\dot{Q} = \frac{T_1 - T_2}{R_A} = \frac{T_2 - T_3}{R_B} = \frac{k_A A(T_1 - T_2)}{L_A}$$

$$= \frac{-k_B A(T_2 - T_3)}{L_B}$$

$$k_A = 2k_B$$

$$L_A = L_B$$

Substitute and solve.

$$\frac{2k_B A(T_1 - T_2)}{L_B} = \frac{k_B A(T_2 - T_3)}{L_B}$$

$$(2)(T_1 - T_2) = T_2 - T_3$$

$$(2)(80°\text{C} - T_2) = T_2 - 25°\text{C}$$

$$T_2 = 61.7°\text{C} \quad (62°\text{C})$$

The answer is (C).

9. Find r_2.

$$r_2 = r_1 + \text{thickness} = 2 \text{ cm} + 3 \text{ cm} = 5 \text{ cm}$$

The thermal resistance is

$$R = \frac{\ln\frac{r_2}{r_1}}{2\pi kL}$$

$$= \frac{\ln\frac{5 \text{ cm}}{2 \text{ cm}}}{(2\pi)\left(0.04 \frac{\text{W}}{\text{m}\cdot°\text{C}}\right)(1 \text{ m})}$$

$$= 3.6 \text{ °C/W}$$

The answer is (B).

10. The Biot number for a sphere is

$$\text{Bi} = \frac{hV}{kA_s} = \frac{h(\frac{4}{3}\pi r^3)}{k(4\pi r^2)} = \frac{hr}{3k}$$

$$= \frac{\left(25 \ \frac{\text{W}}{\text{m}^2\cdot\text{K}}\right)(4 \text{ cm})}{(3)\left(380 \ \frac{\text{W}}{\text{m}\cdot\text{K}}\right)\left(100 \ \frac{\text{cm}}{\text{m}}\right)}$$

$$= 0.00088$$

Since Bi < 0.1, the lumped capacitance method can be used. The temperature variation of a body with time is

$$T - T_\infty = (T_i - T_\infty)e^{-\beta t} = (T_i - T_\infty)e^{-(hA_s/\rho V c_p)t}$$

The equations for the area and volume of a sphere are

$$A_s = \pi d^2$$
$$V = \tfrac{4}{3}\pi r^3 = \frac{\pi d^3}{6}$$

Solve for t.

$$t = \left(\frac{\rho c_p \left(\frac{\pi d^3}{6}\right)}{h\pi d^2}\right) \ln \frac{T_i - T_\infty}{T - T_\infty}$$

$$= \left(\frac{\rho c_p d}{6h}\right) \ln \frac{T_i - T_\infty}{T - T_\infty}$$

$$= \frac{\left(\left(8933 \ \frac{\text{kg}}{\text{m}^3}\right)\left(410 \ \frac{\text{J}}{\text{kg}\cdot\text{K}}\right)(8 \text{ cm})\right) \ln \frac{400\text{K} - 25\text{K}}{200\text{K} - 25\text{K}}}{(6)\left(25 \ \frac{\text{W}}{\text{m}^2\cdot\text{K}}\right)\left(100 \ \frac{\text{cm}}{\text{m}}\right)\left(60 \ \frac{\text{s}}{\text{min}}\right)}$$

$$= 24.8 \text{ min} \quad (25 \text{ min})$$

The answer is (D).

11. Calculate the perimeter length.

$$P = \pi d = \frac{\pi(1.6 \text{ mm})}{1000 \ \frac{\text{mm}}{\text{m}}} = 5.027 \times 10^{-3} \text{ m}$$

Calculate the cross-sectional area of the fin.

$$A_c = \pi r^2 = \pi\left(\frac{d}{2}\right)^2 = \left(\frac{\pi}{4}\right)d^2$$

$$= \frac{\left(\frac{\pi}{4}\right)(1.6 \text{ mm})^2}{\left(1000 \ \frac{\text{mm}}{\text{m}}\right)^2}$$

$$= 2.011 \times 10^{-6} \text{ m}^2$$

Each side of the fin extends outward for an (essentially) infinite distance, so the corrected length is infinite. Since $\tanh \infty = 1$, $\tan(mL_c) = 1$.

With two fins joined at the middle,

$$\dot{Q} = 2\sqrt{hPkA_c}(T_b - T_\infty)\tanh(mL_c)$$

$$= 2\sqrt{\begin{array}{c}\left(17 \ \frac{\text{W}}{\text{m}^2\cdot\text{K}}\right)(5.027 \times 10^{-3} \text{ m}) \\ \times \left(372.1 \ \frac{\text{W}}{\text{m}\cdot\text{K}}\right)(2.011 \times 10^{-6} \text{ m}^2)\end{array}}$$

$$\times (230\text{K} - 27\text{K})(1)$$

$$= 3.25 \text{ W} \quad (3.3 \text{ W})$$

The answer is (B).

20 Convection

PRACTICE PROBLEMS

1. A counterflow heat exchanger is used to cool 0.1 kg/s of oil from 100°C to 70°C. Cooling water enters the heat exchanger at 30°C and leaves at 70°C. The specific heat of oil is 1.9 kJ/kg·°C, and the overall heat transfer coefficient is 0.32 kW/m²·°C. The mean temperature difference correction factor is 0.9. The required heat exchanger area is most nearly

(A) 0.1 m²
(B) 0.3 m²
(C) 0.6 m²
(D) 0.8 m²

2. A white, uninsulated, rectangular duct passes through a room. The duct is 45 cm wide and 30 cm high. Air at 40°C enters the duct flowing at 4.0 m/s. For air at 40°C,

$$\mu = 1.91 \times 10^{-5} \text{ kg/s·m}$$
$$\rho = 1.130 \text{ kg/m}^3$$
$$c_p = 1.0051 \text{ kJ/kg·K}$$
$$k = 0.02718 \text{ W/m·K}$$

What is most nearly the Reynolds number of the air entering the duct?

(A) 6.5×10^4
(B) 7.0×10^4
(C) 8.5×10^4
(D) 9.0×10^4

3. What dimensionless number does the combination of variables shown represent?

$$\frac{c_p \mu}{k}$$

(A) Reynolds
(B) Prandtl
(C) Nusselt
(D) Grashof

4. The transition between laminar and turbulent flow for smooth tubes usually occurs at a Reynolds number of approximately

(A) 900
(B) 1200
(C) 1500
(D) 2300

5. A bare, horizontal conductor with a circular cross section and an outside diameter of 1.5 cm dissipates 25 W per meter of wire length. The conductor is cooled by free convection, and the surrounding air temperature is 15°C. The film temperature is 38°C. The film coefficient is 10.27 W/m²·K. The conductor's surface temperature is most nearly

(A) 29°C
(B) 43°C
(C) 54°C
(D) 67°C

6. 10 kg/s of oil must be cooled from 120°C to 50°C in a counterflow heat exchanger. Water at a temperature of 15°C with a flow rate of 8 kg/s is available. The specific heats of oil and water are 2 kJ/kg·°C and 4.18 kJ/kg·°C, respectively. The process is adiabatic. If the overall heat transfer coefficient is 0.8 kW/m²·°C, the heat transfer area is most nearly

(A) 22 m²
(B) 28 m²
(C) 37 m²
(D) 51 m²

7. A fluid in a tank is maintained at 30°C by an immersed hot water coil. The hot water enters at 90°C and leaves at 70°C. The logarithmic mean temperature difference is most nearly

(A) 49°C
(B) 53°C
(C) 62°C
(D) 71°C

8. An electrically heated plate is mounted vertically in 25°C air. The plate has a surface area of 0.1 m², has a height of 0.3 m, and is maintained at a uniform temperature of 130°C. The average kinematic viscosity of air is 20.92×10^{-6} m²/s, the thermal conductivity is 30×10^{-3} W/m·K, and the Prandtl number is 0.7. The power dissipation is most nearly

(A) 45 W

(B) 60 W

(C) 66 W

(D) 74 W

9. A large horizontal flat heater at 110°C is used to boil water. The density of the water vapor is 0.826 kg/m³, and the density of the liquid water is 951 kg/m³. The surface tension between the water vapor and the liquid water is 0.057 N/m. What is most nearly the heat flux from the heater?

(A) 9.5 kW/m²

(B) 14 kW/m²

(C) 21 kW/m²

(D) 26 kW/m²

SOLUTIONS

1. The heat transfer is

$$\dot{Q} = \dot{m}_{\text{oil}} c_{p,\text{oil}} \Delta T_{\text{oil}}$$
$$= \left(0.1 \; \frac{\text{kg}}{\text{s}}\right)\left(1.9 \; \frac{\text{kJ}}{\text{kg·°C}}\right)(100°\text{C} - 70°\text{C})$$
$$= 5.7 \; \text{kW}$$

The logarithmic mean temperature difference is

$$\Delta T_{\text{lm}} = \frac{(T_{\text{Ho}} - T_{\text{Ci}}) - (T_{\text{Hi}} - T_{\text{Co}})}{\ln \dfrac{T_{\text{Ho}} - T_{\text{Ci}}}{T_{\text{Hi}} - T_{\text{Co}}}}$$

$$= \frac{(70°\text{C} - 30°\text{C}) - (100°\text{C} - 70°\text{C})}{\ln \dfrac{70°\text{C} - 30°\text{C}}{100°\text{C} - 70°\text{C}}}$$

$$= 34.76°\text{C}$$

$$\dot{Q} = UAF\Delta T_{\text{lm}}$$

$$5.7 \; \text{kW} = \left(0.32 \; \frac{\text{kW}}{\text{m}^2 \cdot °\text{C}}\right) A (0.9)(34.76°\text{C})$$

$$A = 0.569 \; \text{m}^2 \quad (0.6 \; \text{m}^2)$$

The answer is (C).

2. The duct is rectangular. The hydraulic diameter is

$$D_H = \frac{4 \times \text{cross-sectional area}}{\text{wetted perimeter}} = \frac{4WH}{2(W+H)}$$

$$= \frac{(4)(45 \; \text{cm})(30 \; \text{cm})}{(2)(45 \; \text{cm} + 30 \; \text{cm})\left(100 \; \dfrac{\text{cm}}{\text{m}}\right)}$$

$$= 0.36 \; \text{m}$$

The Reynolds number is

$$\text{Re} = \frac{\rho v D}{\mu} = \frac{\left(1.130 \; \dfrac{\text{kg}}{\text{m}^3}\right)\left(4.0 \; \dfrac{\text{m}}{\text{s}}\right)(0.36 \; \text{m})}{1.91 \times 10^{-5} \; \dfrac{\text{kg}}{\text{s·m}}}$$

$$= 8.52 \times 10^4 \quad (8.5 \times 10^4)$$

The answer is (C).

3. The equation calculates the Prandtl number.

$$\text{Pr} = \frac{c_p \mu}{k}$$

The answer is (B).

4. Laminar flow in smooth tubes occurs at Reynolds numbers less than 2300.

The answer is (D).

5. The wire diameter is

$$d = \frac{1.5 \text{ cm}}{100 \frac{\text{cm}}{\text{m}}} = 0.015 \text{ m}$$

The heat transfer from the wire is

$$\dot{Q} = hA(T_s - T_\infty) = h\pi d L(T_s - T_\infty)$$

The surface temperature is

$$T_s = \frac{\dfrac{\dot{Q}}{L}}{h\pi d} + T_\infty$$

$$= \frac{\dfrac{25 \text{ W}}{1 \text{ m}}}{\left(10.27 \, \dfrac{\text{W}}{\text{m}^2 \cdot \text{K}}\right)\pi(0.015 \text{ m})} + 15°\text{C}$$

$$= 66.7°\text{C} \quad (67°\text{C})$$

The answer is (D).

6. The heat transfer from the hot oil is

$$\dot{Q}_H = \dot{m}_H c_{p,H}(T_{Hi} - T_{Ho})$$

$$= \left(10 \, \frac{\text{kg}}{\text{s}}\right)\left(2 \, \frac{\text{kJ}}{\text{kg}\cdot°\text{C}}\right)(120°\text{C} - 50°\text{C})$$

$$= 1400 \text{ kW}$$

Since the process is adiabatic, this is also the heat gain by the cold water.

$$\dot{Q}_H = \dot{Q}_C = \dot{m}_C c_{p,C}(T_{Co} - T_{Ci})$$

$$1400 \text{ kW} = \left(8 \, \frac{\text{kg}}{\text{s}}\right)\left(4.18 \, \frac{\text{kJ}}{\text{kg}\cdot°\text{C}}\right)(T_{Co} - 15°\text{C})$$

$$T_{Co} = 56.87°\text{C}$$

$$\Delta T_{\text{lm}} = \frac{(T_{Ho} - T_{Ci}) - (T_{Hi} - T_{Co})}{\ln \dfrac{T_{Ho} - T_{Ci}}{T_{Hi} - T_{Co}}}$$

$$= \frac{(50°\text{C} - 15°\text{C}) - (120°\text{C} - 56.87°\text{C})}{\ln \dfrac{50°\text{C} - 15°\text{C}}{120°\text{C} - 56.87°\text{C}}}$$

$$= 47.69°\text{C}$$

Solve for area.

$$\dot{Q} = UAF\Delta T_{\text{lm}}$$

$$A = \frac{\dot{Q}}{UF\Delta T_{\text{lm}}}$$

$$= \frac{1400 \text{ kW}}{\left(0.8 \, \dfrac{\text{kW}}{\text{m}^2 \cdot °\text{C}}\right)(1.0)(47.69°\text{C})}$$

$$= 36.69 \text{ m}^2 \quad (37 \text{ m}^2)$$

The answer is (C).

7. The logarithmic mean temperature difference is

$$\Delta T_{\text{lm}} = \frac{(T_{Ho} - T_{Co}) - (T_{Hi} - T_{Ci})}{\ln \dfrac{T_{Ho} - T_{Co}}{T_{Hi} - T_{Ci}}}$$

$$= \frac{(70°\text{C} - 30°\text{C}) - (90°\text{C} - 30°\text{C})}{\ln \dfrac{70°\text{C} - 30°\text{C}}{90°\text{C} - 30°\text{C}}}$$

$$= 49.3°\text{C} \quad (49°\text{C})$$

The answer is (A).

8. The absolute surface and air temperatures are

$$T_s = 130°\text{C} + 273°$$

$$= 403\text{K}$$

$$T_\infty = 25°\text{C} + 273°$$

$$= 298\text{K}$$

The coefficient of volumetric expansion is the reciprocal of the absolute temperature of the film, which is the average of the surface and local temperatures.

$$\beta = \frac{2}{T_s + T_\infty}$$

$$= \frac{2}{403\text{K} + 298\text{K}}$$

$$= 0.00285\text{K}^{-1}$$

The Rayleigh number is

$$\text{Ra} = \frac{g\beta(T_s - T_\infty)L^3}{\nu^2}\text{Pr}$$

$$= \left(\frac{\left(9.81\ \frac{\text{m}}{\text{s}^2}\right)\left(0.00285\ \frac{1}{\text{K}}\right)}{\left(20.92 \times 10^{-6}\ \frac{\text{m}^2}{\text{s}}\right)^2}\right)(0.7)$$

$$= 1.27 \times 10^8$$

At this value of Ra, from tabulated values, $C = 0.59$, and $n = 1/4$. The heat transfer coefficient is

$$\bar{h} = C\left(\frac{k}{L}\right)\text{Ra}_L^n$$

$$= (0.59)\left(\frac{30 \times 10^{-3}\ \frac{\text{W}}{\text{m}\cdot\text{K}}}{0.3\ \text{m}}\right)(1.27 \times 10^8)^{1/4}$$

$$= 6.26\ \text{W/m}^2\cdot\text{K}$$

$$\dot{Q} = hA\Delta T$$

$$= \left(6.26\ \frac{\text{W}}{\text{m}^2\cdot\text{K}}\right)(0.1\ \text{m}^2)(403\text{K} - 298\text{K})$$

$$= 65.8\ \text{W}\quad(66\ \text{W})$$

The answer is (C).

9. From the steam tables, the enthalpy of evaporation for water at 110°C is 2230.2 kJ/kg. Use the Zuber equation for minimum heat flux.

$$\dot{q}_{\min} = 0.09\rho_v h_{fg}\left(\frac{\sigma g(\rho_l - \rho_v)}{(\rho_l + \rho_v)^2}\right)^{1/4}$$

$$= (0.09)\left(0.826\ \frac{\text{kg}}{\text{m}^3}\right)\left(2230.2\ \frac{\text{kJ}}{\text{kg}}\right)$$

$$\times \left(\frac{\left(0.057\ \frac{\text{N}}{\text{m}}\right)\left(9.81\ \frac{\text{m}}{\text{s}^2}\right)}{\left(951\ \frac{\text{kg}}{\text{m}^3} + 0.826\ \frac{\text{kg}}{\text{m}^3}\right)^2}\right)^{1/4}$$

$$= 25.8\ \text{kW/m}^2\quad(26\ \text{kW/m}^2)$$

The answer is (D).

21 Radiation

PRACTICE PROBLEMS

1. Which statement concerning radiant heat transfer is FALSE?

(A) The radiation emitted by a body is proportional to the fourth power of its absolute temperature.

(B) For a body of a particular size and temperature, the maximum energy is emitted by a black body.

(C) For an opaque body, the sum of absorptivity and reflectivity is always equal to 1.0.

(D) Radiant energy cannot travel through a vacuum.

2. A 1200°C, 25 cm radius thin-walled cylinder is located concentrically within a hollow, 750°C, 50 cm radius thin-walled cylinder. Both cylinders behave as black bodies.

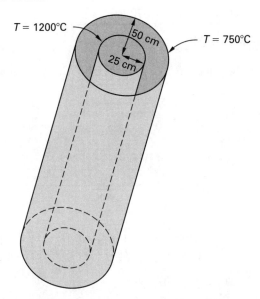

Neglecting conduction, convection, and end effects, what is most nearly the net radiation exchange per unit length between the two cylinders?

(A) 51 kW/m

(B) 160 kW/m

(C) 320 kW/m

(D) 640 kW/m

3. A small tungsten filament (emissivity of 0.35, surface area of 3.0 cm^2) is placed within a spherical enclosure whose temperature is 20°C. What is most nearly the power input needed to keep the filament temperature at 2800K?

(A) 370 W

(B) 480 W

(C) 970 W

(D) 1100 W

4. If T is the absolute temperature, the intensity of radiation from an ideal radiator will be proportional to

(A) T

(B) T^2

(C) T^3

(D) T^4

5. Two plates of equal area are placed parallel to each other in a vacuum. One plate has an emissivity of 0.2 and temperature of 700K, and the other plate has an emissivity of 0.4 and a temperature of 500K. The view factor is 0.8. The radiation heat flux is most nearly

(A) 1.0 kW/m^2

(B) 1.5 kW/m^2

(C) 2.0 kW/m^2

(D) 2.5 kW/m^2

6. A 3 m × 3 m plate at 500°C is suspended vertically in a large room. The plate has an emissivity of 0.13. The room is at 25°C. What is most nearly the net radiant heat transfer from the plate?

(A) 8.3 kW

(B) 24 kW

(C) 46 kW

(D) 47 kW

Problems

7. At 3:00 p.m. in July, a stone wall reaches a maximum temperature of 50°C. The stone has an emissivity of 0.95. Most nearly, what is the radiant heat energy transfer per square meter of the wall?

(A) 190 W/m²
(B) 590 W/m²
(C) 1100 W/m²
(D) 3800 W/m²

8. A copper sphere 0.3 m in diameter radiates power through a vacuum to an environment at 273K. The emissivity of copper is 0.15, and the sphere is at 60°C. The radiation emitted is most nearly

(A) 28 W
(B) 30 W
(C) 37 W
(D) 53 W

9. A metal ball with a radius of 2.5 cm is heated to 573K. The emissivity is 0.4. Most nearly, what is the rate of energy emitted from the surface of the ball?

(A) 0.020 W
(B) 0.70 W
(C) 1.4 W
(D) 19 W

10. In 1 hour, approximately how much black-body radiation escapes a 1 cm × 2 cm rectangular opening in a kiln whose internal temperature is 980°C?

(A) 20 kJ
(B) 100 kJ
(C) 130 kJ
(D) 150 kJ

SOLUTIONS

1. All forms of radiation travel through a vacuum, so the statement "radiant energy cannot travel through a vacuum" is false.

The answer is (D).

2. The net radiation exchange is

$$\dot{Q}_{12} = A_1 F_{12} \sigma (T_1^4 - T_2^4)$$
$$= 2\pi r_1 L F_{12} \sigma (T_1^4 - T_2^4)$$
$$\frac{\dot{Q}_{12}}{L} = 2\pi r_1 F_{12} \sigma (T_1^4 - T_2^4)$$

The shape factor, F_{12}, for a black body enclosed completely within another black body is equal to 1.

$$\frac{\dot{Q}_{12}}{L} = 2\pi r_1 F_{12} \sigma (T_1^4 - T_2^4)$$

$$= \frac{(2\pi)(0.25 \text{ m})(1)\left(5.67 \times 10^{-8} \, \frac{\text{W}}{\text{m}^2 \cdot \text{K}^4}\right)}{1000 \, \frac{\text{W}}{\text{kW}}} \times ((1200°\text{C} + 273°)^4 - (750°\text{C} + 273°)^4)$$

$$= 321.7 \text{ kW/m} \quad (320 \text{ kW/m})$$

The answer is (C).

3. Let the filament be body 1, and the enclosure be body 2.

$$T_2 = 20°\text{C} + 273° = 293\text{K}$$
$$\dot{Q}_{12} = F_{12} \sigma A_1 (T_1^4 - T_2^4)$$
$$= \varepsilon_1 \sigma A_1 (T_1^4 - T_2^4)$$
$$= (0.35)\left(5.67 \times 10^{-8} \, \frac{\text{W}}{\text{m}^2 \cdot \text{K}^4}\right)$$
$$\times \left(\frac{3.0 \text{ cm}^2}{\left(100 \, \frac{\text{cm}}{\text{m}}\right)^2}\right) \left((2800\text{K})^4 - (293\text{K})^4\right)$$
$$= 365.9 \text{ W} \quad (370 \text{ W})$$

The answer is (A).

4. The intensity of an ideal (black-body) radiator at absolute temperature, T, is given by the Stefan-Boltzmann law, also known as the fourth power law.

$$\dot{Q}_{\text{black}} = \varepsilon \sigma A T^4$$

The answer is (D).

5. The net heat transfer due to radiation is

$$\dot{Q}_{12} = \frac{\sigma(T_1^4 - T_2^4)}{\frac{1-\varepsilon_1}{\varepsilon_1 A_1} + \frac{1}{A_1 F_{12}} + \frac{1-\varepsilon_2}{\varepsilon_2 A_2}}$$

Solve for heat flux.

$$\frac{\dot{Q}_{12}}{A} = \frac{\sigma(T_1^4 - T_2^4)}{\frac{1-\varepsilon_1}{\varepsilon_1} + \frac{1}{F_{12}} + \frac{1-\varepsilon_2}{\varepsilon_2}}$$

$$= \frac{\left(5.67 \times 10^{-8} \ \frac{W}{m^2 \cdot K^4}\right) \times \left((700K)^4 - (500K)^4\right)}{\left(\frac{1-0.2}{0.2} + \frac{1}{0.8} + \frac{1-0.4}{0.4}\right)\left(1000 \ \frac{W}{kW}\right)}$$

$$= 1.492 \ kW/m^2 \quad (1.5 \ kW/m^2)$$

The answer is (B).

6. The plate has two surfaces. Disregarding the edges, the radiating area is

$$A = (2)(3 \ m)(3 \ m) = 18 \ m^2$$

The configuration factor for a fully enclosed body is the emissivity of the body. In this case,

$$F_{12} = \varepsilon_1 = 0.13$$
$$\dot{Q}_{12} = A_1 F_{12} \sigma(T_1^4 - T_2^4)$$

$$= \frac{(18 \ m^2)(0.13)\left(5.67 \times 10^{-8} \ \frac{W}{m^2 \cdot K^4}\right) \times \left((500°C + 273°)^4 - (25°C + 273°)^4\right)}{1000 \ \frac{W}{kW}}$$

$$= 46.3 \ kW \quad (46 \ kW)$$

The answer is (C).

7. The energy radiated by a gray body is

$$\dot{Q} = \varepsilon \sigma A T^4$$

The heat flux is

$$\dot{q} = \frac{\dot{Q}}{A} = \varepsilon \sigma T^4$$

$$= (0.95)\left(5.67 \times 10^{-8} \ \frac{W}{m^2 \cdot K^4}\right)(50°C + 273°)^4$$

$$= 586 \ W/m^2 \quad (590 \ W/m^2)$$

The answer is (B).

8. The rate of energy emitted is

$$\dot{Q} = \varepsilon \sigma A T^4 = \varepsilon \sigma 4\pi r^2 T^4$$

$$= (0.15)\left(5.67 \times 10^{-8} \ \frac{W}{m^2 \cdot K^4}\right)$$
$$\times (4\pi)\left(\frac{0.3 \ m}{2}\right)^2 (60°C + 273°)^4$$

$$= 29.6 \ W \quad (30 \ W)$$

The answer is (B).

9. The rate of energy emitted is

$$Q = \varepsilon \sigma A T^4 = \varepsilon \sigma 4\pi r^2 T^4$$

$$= (0.4)\left(5.67 \times 10^{-8} \ \frac{W}{m^2 \cdot K^4}\right)$$
$$\times (4\pi)\left(\frac{2.5 \ cm}{100 \ \frac{cm}{m}}\right)^2 (573K)^4$$

$$= 19.2 \ W \quad (19 \ W)$$

The answer is (D).

10. Calculate the area of the opening.

$$A = (1 \ cm)(2 \ cm)$$
$$= 2 \ cm^2$$

The black-body radiation is

$$\dot{Q}_{black} = \varepsilon \sigma A T^4$$

$$= \frac{(1)\left(5.67 \times 10^{-8} \ \frac{W}{m^2 \cdot K^4}\right)(2 \ cm^2) \times (980°C + 273°)^4}{\left(100 \ \frac{cm}{m}\right)^2}$$

$$= 28 \ W$$

$$Q = \dot{Q} t$$

$$= \frac{(28 \ W)(1 \ h)\left(3600 \ \frac{s}{h}\right)}{1000 \ \frac{J}{kJ}}$$

$$= 100.6 \ kJ \quad (100 \ kJ)$$

The answer is (B).

22 Systems of Forces and Moments

PRACTICE PROBLEMS

1. In the system shown, force F, line P, and line Q are coplanar.

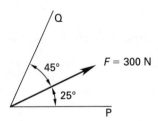

Resolve the 300 N force into two components, one along line P and the other along line Q.

(A) $F_Q = 272$ N; $F_P = 126$ N
(B) $F_Q = 232$ N; $F_P = 186$ N
(C) $F_Q = 135$ N; $F_P = 226$ N
(D) $F_Q = 212$ N; $F_P = 226$ N

2. Which type of load is NOT resisted by a pinned joint?

(A) moment
(B) shear
(C) axial
(D) distributed

3. The loading shown requires a clockwise resisting moment of 20 N·m at the support.

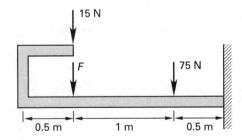

What is most nearly the value of force F?

(A) 25 N (up)
(B) 27 N (up)
(C) 38 N (down)
(D) 43 N (down)

4. A bent beam is acted upon by a moment and several concentrated forces, as shown.

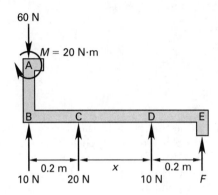

Approximate the unknown force, F, and distance, x, that will maintain equilibrium on the beam.

(A) $F = 5$ N; $x = 0.8$ m
(B) $F = 10$ N; $x = 0.6$ m
(C) $F = 20$ N; $x = 0.2$ m
(D) $F = 20$ N; $x = 0.4$ m

5. For the member shown, the 700 N·m moment is applied at point B.

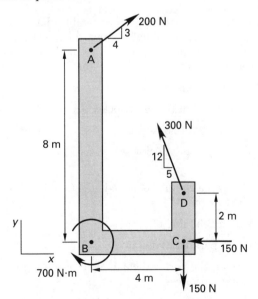

If the member is pinned so that it rotates around point B, what counteracting moment must be applied at point A to keep the member in equilibrium?

(A) 650 N·m
(B) 890 N·m
(C) 1150 N·m
(D) 1240 N·m

6. A force is defined by the vector $\mathbf{A} = 3.5\mathbf{i} - 1.5\mathbf{j} + 2.0\mathbf{k}$. What is most nearly the angle between the force and the positive y-axis?

(A) 20°
(B) 66°
(C) 70°
(D) 110°

7. In the structure shown, the beam is pinned at point B. Point E is a roller support. The beam is loaded with a distributed load from point A to point B of 400 N/m, a 500 N·m couple at point C, and a vertical 900 N force at point D.

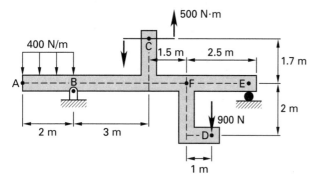

If the distributed load and the vertical load are removed and replaced with a vertically upward force of 1700 N at point F, what moment at point F would be necessary to keep the reaction at point E the same?

(A) −9000 N·m (counterclockwise)
(B) −6500 N·m (counterclockwise)
(C) 3500 N·m (clockwise)
(D) 12 000 N·m (clockwise)

8. Where can a couple be moved on a rigid body to have an equivalent effect?

(A) along the line of action
(B) in a parallel plane
(C) along the perpendicular bisector joining the two original forces
(D) anywhere on the rigid body

9. The overhanging beam shown is supported by a roller and a pinned support. The moment is removed and replaced by a couple consisting of forces applied at points A and C.

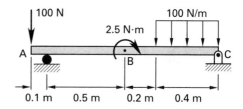

What is the magnitude of the forces that constitute the couple?

(A) 2.1 N
(B) 4.2 N
(C) 6.3 N
(D) 8.3 N

10. A disk-shaped body with a 4 cm radius has a 320 N force acting through the center at an unknown angle θ, and two 40 N loads acting as a couple, as shown. All of these forces are removed and replaced by a single 320 N force at point B, parallel to the original 320 N force.

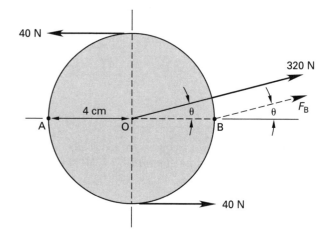

Most nearly, what is the angle θ?

(A) 0°
(B) 7.6°
(C) 15°
(D) 29°

11. The overhanging beam shown is supported by a roller and a pinned support. The moment is removed and replaced by a couple consisting of forces applied at points A and C.

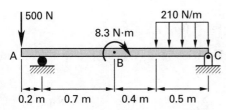

What is most nearly the magnitude of the couple that exactly replaces the moment that is removed?

(A) 0.080 N·m

(B) 0.16 N·m

(C) 8.3 N·m

(D) 15 N·m

12. Which structure is statically determinate and stable with the loadings shown?

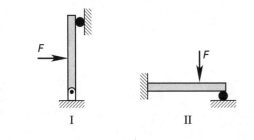

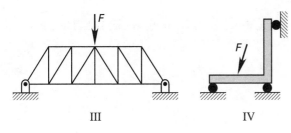

(A) I only

(B) I and III

(C) I and IV

(D) II and III

13. A signal arm carries two traffic signals and a sign, as shown. The signals and sign are rigidly attached to the arm. Each traffic signal has a frontal area of 0.2 m² and weighs 210 N. The sign weighs 60 N per square meter of area. The design wind pressure is 575 Pa. The maximum moment that the connection between the arm and pole can withstand due to wind is 6000 N·m, and the maximum permitted moment due to the loads is 4000 N·m.

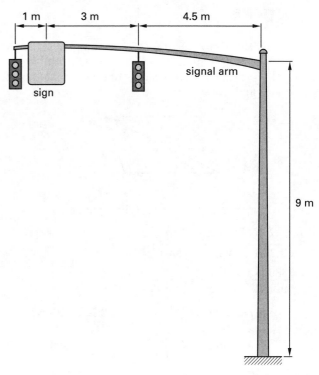

As limited by moment on the connection, what is most nearly the maximum area of the sign?

(A) 1.0 m²

(B) 1.2 m²

(C) 2.8 m²

(D) 5.6 m²

14. A hinged arch is composed of two pin-connected curved members supported on two pinned supports, as shown. Both members are rigid. A horizontal force of 1000 N is applied to pin B, as shown. All coordinates are in meters.

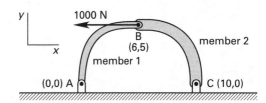

Most nearly, what are the reactions and moments at joint A?

(A) $F_x = 500$ N; $F_y = 600$ N; $M_A = 5000$ N·m

(B) $F_x = 600$ N; $F_y = 500$ N; $M_A = 0$ N·m

(C) $F_x = 680$ N; $F_y = 400$ N; $M_A = 5000$ N·m

(D) $F_x = 616$ N; $F_y = 480$ N; $M_A = 0$ N·m

SOLUTIONS

1. One of the characteristics of the components of a force is that they combine as vectors into the total force. Draw the vector addition triangle and determine all of the angles and sides.

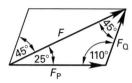

Use the law of sines to calculate the components.

$$\frac{F}{\sin 110°} = \frac{F_Q}{\sin 25°} = \frac{F_P}{\sin 45°}$$

$$\frac{300 \text{ N}}{\sin 110°} = \frac{F_Q}{\sin 25°} = \frac{F_P}{\sin 45°}$$

$$F_Q = (300 \text{ N})\frac{\sin 25°}{\sin 110°} = 134.9 \text{ N} \quad (135 \text{ N})$$

$$F_P = (300 \text{ N})\frac{\sin 45°}{\sin 110°} = 225.7 \text{ N} \quad (226 \text{ N})$$

The answer is (C).

2. A pinned support will resist forces but not moments.

The answer is (A).

3. Clockwise moments are positive. The sum of the moments around the support is

$$\sum M = 0$$
$$= 20 \text{ N·m} - (75 \text{ N})(0.5 \text{ m}) - F(1.5 \text{ m})$$
$$\quad - (15 \text{ N})(1.5 \text{ m})$$
$$F = -26.7 \text{ N} \quad (27 \text{ N}) \quad [\text{up}]$$

The answer is (B).

4. The sum of the forces in the y-direction is

$$\sum F_y = 0$$
$$= -60 \text{ N} + 10 \text{ N} + 20 \text{ N} + 10 \text{ N} + F$$
$$F = 20 \text{ N}$$

Clockwise moments are positive. The sum of the moments around point A is

$$\sum M_A = 0$$
$$= 20 \text{ N·m} - (20 \text{ N})(0.2 \text{ m})$$
$$\quad - (10 \text{ N})(0.2 \text{ m} + x) - (20 \text{ N})(0.4 \text{ m} + x)$$

$$4 + 2 + 10x + 8 + 20x = 20$$
$$30x = 6$$
$$x = 0.2 \text{ m}$$

The answer is (C).

5. Let clockwise moments be positive. Take moments about point B.

$$\sum M_B = 700 \text{ N·m} + (150 \text{ N})(4 \text{ m})$$
$$\quad - (300 \text{ N})\left(\frac{5}{13}\right)(2 \text{ m})$$
$$\quad - (300 \text{ N})\left(\frac{12}{13}\right)(4 \text{ m})$$
$$\quad + (200 \text{ N})\left(\frac{4}{5}\right)(8 \text{ m})$$
$$= 1242 \text{ N·m} \quad (1240 \text{ N·m})$$

The application point of the moment is irrelevant.

The answer is (D).

6. Calculate the dot product of two vectors.

$$\mathbf{A} = 3.5\mathbf{i} - 1.5\mathbf{j} + 2\mathbf{k}$$
$$\mathbf{B} = \mathbf{j}$$
$$\mathbf{A} \cdot \mathbf{B} = a_x b_x + a_y b_y + a_z b_z$$
$$= (3.5)(0) + (-1.5)(1) + (2)(0)$$
$$= -1.5$$

The magnitude of $\mathbf{B}$ (a unit vector) is 1. The magnitude of $\mathbf{A}$ (same as F) is

$$F = \sqrt{F_x^2 + F_y^2 + F_z^2}$$
$$= \sqrt{(3.5)^2 + (-1.5)^2 + (2.0)^2}$$
$$= 4.3$$

The dot product is defined as the product of the vector magnitudes multiplied by the cosine of the angle between them.

$$\mathbf{A} \cdot \mathbf{B} = |\mathbf{A}||\mathbf{B}| \cos \theta$$
$$-1.5 = (4.3)(1)\cos \theta$$
$$\theta = \arccos \frac{-1.5}{4.3} = 110.4° \quad (110°)$$

The answer is (D).

7. The reaction at point E is unknown, but it is irrelevant. Since the reaction is to be unchanged, it is necessary only to calculate the change in the loading.

Assume clockwise moments are positive. Take moments about point B for the forces that are removed and added.

$$\sum M_B = \sum M_{B,\text{removed}} - \sum M_{B,\text{added}}$$
$$= -\left(400 \ \frac{N}{m}\right)(2 \text{ m})\left(\frac{2 \text{ m}}{2}\right)$$
$$+ (900 \text{ N})(1 \text{ m} + 1.5 \text{ m} + 3 \text{ m})$$
$$- (-1700 \text{ N})(1.5 \text{ m} + 3 \text{ m})$$
$$= 11\,800 \text{ N·m} \quad (12\,000 \text{ N·m}) \quad [\text{clockwise}]$$

This is the moment that is applied by the forces that are removed, reduced by the moment of the new force that is applied. A 12 000 N·m clockwise moment must be applied to counteract the change. The location of the new moment is not relevant.

The answer is (D).

8. Since a couple is composed of two equal but opposite forces, the x- and y-components will always cancel, no matter what the orientation. Only the moment produced by the couple remains.

The answer is (D).

9. The distance between the forces is

$$d = 0.4 \text{ m} + 0.2 \text{ m} + 0.5 \text{ m} + 0.1 \text{ m}$$
$$= 1.2 \text{ m}$$

The forces are

$$F = \frac{M}{d} = \frac{2.5 \text{ N·m}}{1.2 \text{ m}}$$
$$= 2.08 \text{ N} \quad (2.1 \text{ N})$$

The answer is (A).

10. Assume clockwise moments are positive. Take moments about the center for the original forces. The 320 N force has no moment arm, so it does not contribute to the moment. The couple is

$$M = Fd = -(40 \text{ N})(8 \text{ cm})$$
$$= -320 \text{ N·cm}$$

The replacement force must produce a moment of -320 N·cm. The horizontal component of the replacement force acts through the center; only the vertical component of the force contributes to the moment.

$$M = -320 \text{ N·cm} = Fr = -(320 \text{ N})(\sin\theta)(4 \text{ cm})$$
$$\theta = 14.48° \quad (15°)$$

The answer is (C).

11. A couple is a moment. When a moment of 8.3 N·m is removed, it must be replaced by the same moment.

The answer is (C).

12. Structure I is simply supported and determinate. Structure II is a propped cantilever beam, always indeterminate by one degree. Structure III is a truss that is pinned at both ends, also indeterminate by one degree. Structure IV is a beam with three rollers, two in the vertical direction and one in the horizontal direction. It is determinate, but not stable.

The answer is (A).

13. The length of the signal arm is

$$1 \text{ m} + 3 \text{ m} + 4.5 \text{ m} = 8.5 \text{ m}$$

Set the moment on the arm due to the wind equal to the maximum allowed.

$$\sum M_{\text{wind}} = (0.2 \text{ m}^2)(575 \text{ Pa})(8.5 \text{ m} + 4.5 \text{ m})$$
$$+ A_{\text{sign}}(575 \text{ Pa})(7.5 \text{ m}) = 6000 \text{ N·m}$$
$$A_{\text{sign}} = 1.04 \text{ m}^2$$

Set the moment on the arm due to vertical loading equal to the maximum allowed.

$$\sum M_{\text{loads}} = (210 \text{ N})(8.5 \text{ m} + 4.5 \text{ m})$$
$$+ \left(60 \ \frac{N}{m^2}\right) A_{\text{sign}}(7.5 \text{ m}) = 4000 \text{ N·m}$$
$$A_{\text{sign}} = 2.82 \text{ m}^2$$

The maximum area of the sign is the smaller of these two values, 1.04 m² (1.0 m²).

The answer is (A).

14. Sum moments about point C. The x-component of the reaction at point A has a zero moment arm.

$$\sum M_C = 10F_y - (1000 \text{ N})(5) = 0$$
$$F_y = 500 \text{ N}$$

Point B is a pin, which transmits no moment. Sum moments to the left of point B.

$$\sum M_B = F_y(6) - F_x(5) = 0$$
$$= (500 \text{ N})(6) - F_x(5) = 0$$
$$F_x = 600 \text{ N}$$

Point A is a pin, which transmits no moment.

The answer is (B).

23 Trusses

PRACTICE PROBLEMS

1. Determine the approximate force in member BC for the truss shown.

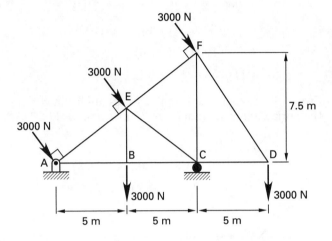

(A) 0 N
(B) 1000 N
(C) 1500 N
(D) 2500 N

2. Find the approximate force in member DE for the truss shown.

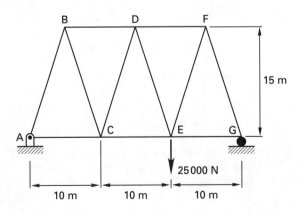

(A) 0 N
(B) 6300 N
(C) 8800 N
(D) 10 000 N

3. For the truss shown, what are most nearly the forces in members AC and BD?

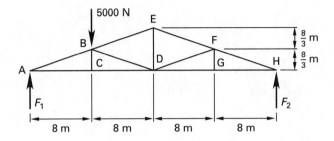

(A) $AC = 11\,000$ N; $BD = -7900$ N
(B) $AC = 0$ N; $BD = -2000$ N
(C) $AC = 1100$ N; $BD = 2500$ N
(D) $AC = 0$ N; $BD = -7900$ N

4. The braced frame shown is constructed with pin-connected members and supports. All applied forces are horizontal.

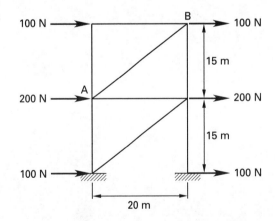

Most nearly, what is the force in the diagonal member BA?

(A) 0 N
(B) 160 N
(C) 200 N
(D) 250 N

5. For the cantilever truss shown, what is most nearly the force in member AF?

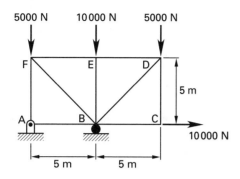

(A) 0 N

(B) 5000 N

(C) 10 000 N

(D) 15 000 N

SOLUTIONS

1. Distance AF is

$$AF = \sqrt{(5 \text{ m} + 5 \text{ m})^2 + (7.5 \text{ m})^2} = 12.5 \text{ m}$$

Distance AE is

$$AE = \tfrac{1}{2} AF = \left(\tfrac{1}{2}\right)(12.5 \text{ m}) = 6.25 \text{ m}$$

The sum of the moments around point A is

$$\sum M_A = 0$$
$$= (3000 \text{ N})(6.25 \text{ m}) + (3000 \text{ N})(12.5 \text{ m})$$
$$\quad + (3000 \text{ N})(5 \text{ m}) - F_{C_y}(10 \text{ m})$$
$$\quad + (3000 \text{ N})(15 \text{ m})$$
$$F_{C_y} = 11\,625 \text{ N} \quad \text{[upward]}$$

From trigonometry, the applied forces are inclined from the horizontal at $\arctan(7.5 \text{ m}/(5 \text{ m} + 5 \text{ m})) = 36.87°$.

$$\sum F_y = 0$$
$$= F_{A_y} - (3)(3000 \text{ N})\cos 36.87°$$
$$\quad - (2)(3000 \text{ N}) + 11\,625 \text{ N}$$
$$F_{A_y} = 1575 \text{ N} \quad \text{[upward]}$$
$$\sum F_x = 0$$
$$= F_{A_x} + (3)(3000 \text{ N})\sin 36.87°$$
$$F_{A_x} = -5400 \text{ N} \quad \text{[to the left]}$$

Use the method of sections.

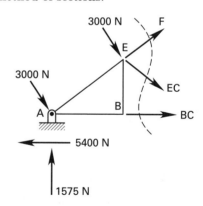

Take moments about point E. The vertical downward force at point B passes through point E, and it does not generate a moment.

$$\sum M_E = 0$$
$$= (5400 \text{ N})(3.75 \text{ m}) + (1575 \text{ N})(5 \text{ m})$$
$$\quad - (3000 \text{ N})(6.25 \text{ m}) - BC(3.75 \text{ m})$$
$$BC = 2500 \text{ N}$$

The answer is (D).

2. Take moments about point G.

$$\sum M_G = 0$$
$$= (-25\,000 \text{ N})(10 \text{ m}) + F_{A_y}(30 \text{ m})$$
$$F_{A_y} = 8333 \text{ N} \quad \text{[upward]}$$

Use the method of sections.

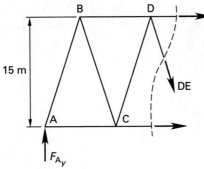

$$\sum F_y = 0$$
$$= 8333 \text{ N} - DE_y$$
$$DE_y = 8333 \text{ N}$$
$$DE_x = (8333 \text{ N})\left(\frac{5 \text{ m}}{15 \text{ m}}\right)$$
$$= 2778 \text{ N}$$
$$DE = \sqrt{(8333 \text{ N})^2 + (2778 \text{ N})^2}$$
$$= 8784 \text{ N} \quad (8800 \text{ N}) \quad \text{[tension]}$$

The answer is (C).

3. Take moments about point H.

$$\sum M_H = 0$$
$$= (5000 \text{ N})(24 \text{ m}) - F_1(32 \text{ m})$$
$$F_1 = 3750 \text{ N}$$

The angle made by the inclined members with the horizontal is

$$\arctan \frac{\tfrac{8}{3} \text{ m}}{8 \text{ m}} = 18.435°$$

(Alternatively, the force components could be determined from geometry.)

Use the method of joints.

For pin A,

$$\sum F_y = 0$$
$$= F_1 + AB \sin 18.435°$$
$$= 3750 \text{ N} + AB \sin 18.435°$$
$$AB = -11\,859 \text{ N} \quad \text{[compression]}$$
$$\sum F_x = 0$$
$$= (-11\,859 \text{ N})\cos 18.435° + AC$$
$$AC = 11\,250 \text{ N} \quad (11\,000 \text{ N}) \quad \text{[tension]}$$

For pin C,

$$\sum F_y = 0$$
$$BC = 0 \quad \text{[zero-force member]}$$

For pin B,

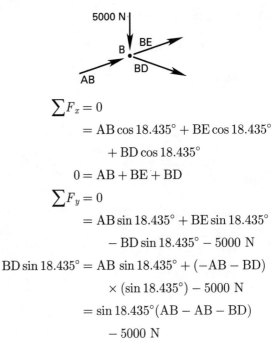

$$\sum F_x = 0$$
$$= AB \cos 18.435° + BE \cos 18.435°$$
$$\quad + BD \cos 18.435°$$
$$0 = AB + BE + BD$$
$$\sum F_y = 0$$
$$= AB \sin 18.435° + BE \sin 18.435°$$
$$\quad - BD \sin 18.435° - 5000 \text{ N}$$
$$BD \sin 18.435° = AB \sin 18.435° + (-AB - BD)$$
$$\quad \times (\sin 18.435°) - 5000 \text{ N}$$
$$= \sin 18.435°(AB - AB - BD)$$
$$\quad - 5000 \text{ N}$$
$$2BD \sin 18.435° = -5000 \text{ N}$$
$$BD = -7906 \text{ N} \quad (-7900 \text{ N})$$
$$\begin{bmatrix} \text{compression in opposite} \\ \text{direction shown} \end{bmatrix}$$

The answer is (A).

4. Determine the length of member BA by recognizing this configuration to be a 3-4-5 triangle.

$$L_{BA} = 25 \text{ m}$$

Use the method of sections. Cut the frame horizontally through member BA.

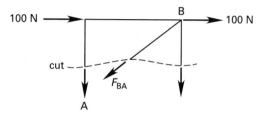

By inspection, the horizontal component of F_{BA} balances the two applied horizontal loads.

$$BA_x = 100 \text{ N} + 100 \text{ N} = 200 \text{ N}$$

By similar triangles,

$$BA = \left(\tfrac{5}{4}\right)(200 \text{ N}) = 250 \text{ N}$$

The answer is (D).

5. Take moments about point B.

$$\sum M_B = 0$$
$$= -(5000 \text{ N})(5 \text{ m}) + (5000 \text{ N})(5 \text{ m})$$
$$\quad + R_{A_y}(5 \text{ m})$$
$$R_{A_y} = 0 \text{ N}$$

By inspection, the force in member AF is equal to the vertical component of the reaction at point A.

The answer is (A).

24 Pulleys, Cables, and Friction

PRACTICE PROBLEMS

1. A 2 kg block rests on a 34° incline.

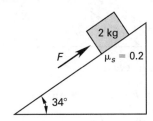

If the coefficient of static friction is 0.2, approximately how much additional force, F, must be applied to keep the block from sliding down the incline?

(A) 7.7 N
(B) 8.8 N
(C) 9.1 N
(D) 14 N

2. A box has uniform density and a total weight of 600 N. It is suspended by three equal-length cables, AE, BE, and CE, as shown. Point E is 0.5 m directly above the center of the box's top surface.

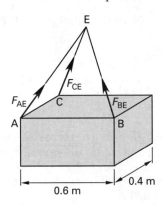

Most nearly, what is the tension in cable CE?

(A) 130 N
(B) 200 N
(C) 370 N
(D) 400 N

3. A rope is wrapped over a 6 cm diameter pipe to support a bucket of tools being lowered. The coefficient of static friction between the rope and the pipe is 0.20. The combined mass of the bucket and tools is 100 kg.

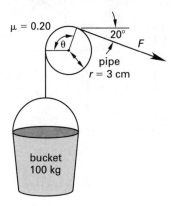

What is most nearly the range of force that can be applied to the free end of the rope such that the bucket remains stationary?

(A) 560 N to 1360 N
(B) 670 N to 1440 N
(C) 720 N to 1360 N
(D) 720 N to 1510 N

4. The two cables shown carry a 100 N vertical load.

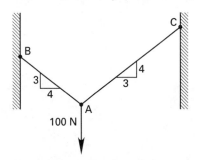

Most nearly, what is the tension in cable AB?

(A) 40 N
(B) 50 N
(C) 60 N
(D) 80 N

5. Most nearly, what is the tension, T, that must be applied to pulley A to lift the 1200 N weight?

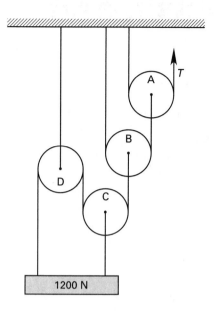

(A) 75 N
(B) 100 N
(C) 300 N
(D) 400 N

SOLUTIONS

1. Choose coordinate axes parallel and perpendicular to the incline.

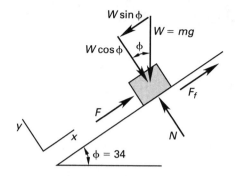

The sum of the forces is

$$\sum F_x = 0 = F + F_f - W\sin\phi$$
$$F = W\sin\phi - F_f$$
$$= mg\sin\phi - \mu_s N$$
$$= mg\sin\phi - \mu_s mg\cos\phi$$
$$= mg(\sin\phi - \mu\cos\phi)$$
$$= (2\text{ kg})\left(9.81\ \frac{\text{m}}{\text{s}^2}\right)(\sin 34° - 0.2\cos 34°)$$
$$= 7.7\text{ N}$$

The answer is (A).

2. The length of the diagonal is

$$\text{BC} = \sqrt{(0.4\text{ m})^2 + (0.6\text{ m})^2} = 0.721\text{ m}$$

The cable length is

$$\text{BE} = \sqrt{\left(\frac{0.721\text{ m}}{2}\right)^2 + (0.5\text{ m})^2} = 0.616\text{ m}$$

There is nothing to balance the component force in the direction from point A to the opposite corner, so the force in cable AE is zero. Cables BE and CE each carry half of the box weight. The vertical component of force in each cable is 300 N.

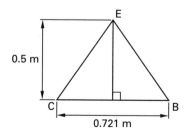

By similar triangles, the tensile force in each cable is

$$T = \frac{(300 \text{ N})(0.616 \text{ m})}{0.5 \text{ m}} = 370 \text{ N}$$

The answer is (C).

3. The angle of wrap is

$$\theta = (90° + 20°)\left(\frac{2\pi \text{ rad}}{360°}\right) = 1.92 \text{ rad}$$

(This must be expressed in radians.)

The tensile force in the rope due to the bucket's mass is

$$F = mg = (100 \text{ kg})\left(9.81 \ \frac{\text{m}}{\text{s}^2}\right)$$
$$= 981 \text{ N}$$

The free end of the rope can either be on the tight or loose side. These two options define the range of force that will keep the bucket stationary.

The ratio of tight-side to loose-side forces is

$$\frac{F_1}{F_2} = e^{\mu\theta} = e^{(0.20)(1.92 \text{ rad})}$$
$$= 1.468$$

Multiply and divide the bucket-end tension by this ratio.

minimum tension: $\quad \frac{981 \text{ N}}{1.468} = 668 \text{ N} \quad (670 \text{ N})$

maximum tension: $\quad (1.468)(981 \text{ N}) = 1440 \text{ N}$

The answer is (B).

4. Recognize that the orientations of both cables are defined by 3-4-5 triangles.

The equilibrium condition for horizontal forces at point A is

$$F_x = T_{AC} - T_{AB} = 0$$
$$\tfrac{3}{5}T_{AC} - \tfrac{4}{5}T_{AB} = 0$$
$$T_{AC} = \tfrac{4}{3}T_{AB}$$

The equilibrium condition for vertical forces at point A is

$$F_y = T_{AB} + T_{AC} - 100 \text{ N} = 0$$
$$\tfrac{3}{5}T_{AB} + \tfrac{4}{5}T_{AC} - 100 \text{ N} = 0$$

Substitute $(4/3)T_{AB}$ for T_{AC}.

$$\tfrac{3}{5}T_{AB} + \left(\tfrac{4}{5}\right)\left(\tfrac{4}{3}\right)T_{AB} = 100 \text{ N}$$
$$\left(\tfrac{3}{5} + \tfrac{16}{15}\right)T_{AB} = 100 \text{ N}$$
$$T_{AB} = 60 \text{ N}$$

The answer is (C).

5. The free bodies of the system are shown.

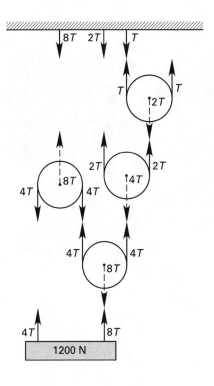

$$\sum F_y = 0$$
$$= -1200 \text{ N} + 4T + 8T$$
$$12T = 1200 \text{ N}$$
$$T = 100 \text{ N}$$

The answer is (B).

25 Centroids and Moments of Inertia

PRACTICE PROBLEMS

1. Refer to the complex shape shown.

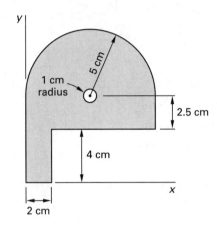

What is most nearly the moment of inertia about the x-axis?

(A) 1500 cm^4

(B) 3400 cm^4

(C) 3600 cm^4

(D) 5200 cm^4

2. Refer to the composite plane areas shown.

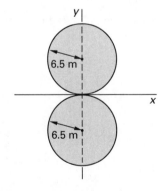

What is the approximate polar moment of inertia about the composite centroid?

(A) 2400 m^4

(B) 5500 m^4

(C) 12 000 m^4

(D) 17 000 m^4

3. Most nearly, what is the area moment of inertia about the x-axis for the area shown?

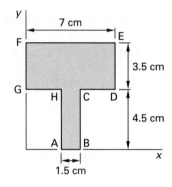

(A) 89 cm^4

(B) 170 cm^4

(C) 510 cm^4

(D) 1000 cm^4

4. Refer to the cross section of the angle shown.

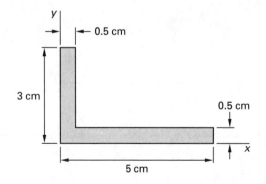

Most nearly, what is the x-coordinate of the centroid of the perimeter line?

(A) 1.56 cm

(B) 1.66 cm

(C) 1.75 cm

(D) 1.80 cm

5. The centroidal moment of inertia about the x-axis for the area shown is 142.41 cm⁴.

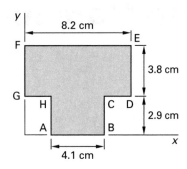

Most nearly, what is the centroidal polar moment of inertia?

(A) 79 cm⁴
(B) 110 cm⁴
(C) 330 cm⁴
(D) 450 cm⁴

6. Refer to the complex shape shown.

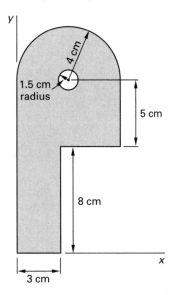

What is most nearly the y-coordinate of the centroid?

(A) 5.5 cm
(B) 7.3 cm
(C) 9.7 cm
(D) 11 cm

7. Most nearly, what are the x- and y-coordinates of the centroid of the perimeter line for the area shown?

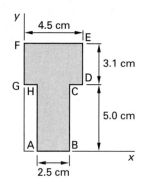

(A) 1.0 cm; 3.8 cm
(B) 1.0 cm; 4.0 cm
(C) 2.3 cm; 4.4 cm
(D) 2.3 cm; 4.8 cm

8. Refer to the composite plane areas shown. The polar moment of inertia is $3\pi r^4$.

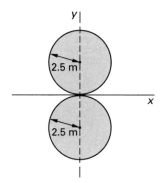

What is most nearly the polar radius of gyration?

(A) 2.5 m
(B) 2.7 m
(C) 2.9 m
(D) 3.1 m

9. The y-coordinate of the centroid of the area shown is 5.8125 cm.

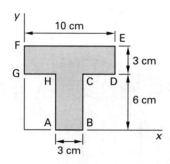

Most nearly, what is the centroidal moment of inertia with respect to the x-axis?

(A) 82 cm^4

(B) 100 cm^4

(C) 220 cm^4

(D) 300 cm^4

10. Refer to the complex shape shown.

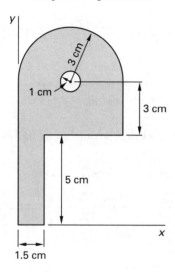

What is most nearly the x-coordinate of the centroid?

(A) 2.4 cm

(B) 2.5 cm

(C) 2.8 cm

(D) 3.2 cm

11. The centroidal moment of inertia of area A_2 with respect to the composite centroidal x-axis is 73.94 cm^4. The y-coordinate of the composite centroid is 4.79 cm. The moment of inertia of area A_3 with respect to the composite centroidal x-axis is 32.47 cm^4.

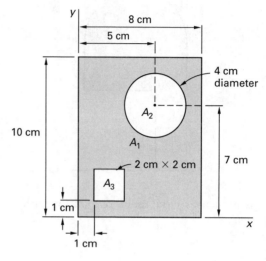

Most nearly, what is the moment of inertia of the entire composite area with respect to its centroidal x-axis?

(A) 350 cm^4

(B) 460 cm^4

(C) 480 cm^4

(D) 560 cm^4

12. Most nearly, what are the x- and y-coordinates of the centroid of the area shown?

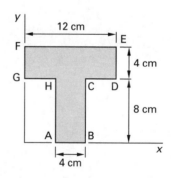

(A) 4.8 cm; 6.8 cm

(B) 6.0 cm; 7.2 cm

(C) 6.0 cm; 7.6 cm

(D) 6.0 cm; 8.0 cm

13. Refer to the complex shape shown. The moment of inertia about the y-axis is 352 cm^4, and the total area of the shape is 36.5 cm^2.

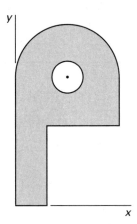

What is most nearly the radius of gyration with respect to the y-axis?

(A) 1.9 cm

(B) 3.1 cm

(C) 3.3 cm

(D) 4.0 cm

SOLUTIONS

1. Calculate the centroidal moment of inertia, the area, and the distance from that centroid to the x-axis for each region.

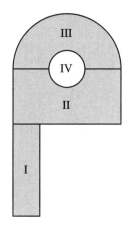

For region I (rectangular),

$$A = (2 \text{ cm})(4 \text{ cm}) = 8 \text{ cm}^2$$

$$I_c = \frac{bh^3}{12} = \frac{(2 \text{ cm})(4 \text{ cm})^3}{12} = 10.67 \text{ cm}^4$$

$$d = \frac{4 \text{ cm}}{2} = 2 \text{ cm}$$

$$d^2 A = (2 \text{ cm})^2 (8 \text{ cm}^2) = 32 \text{ cm}^4$$

For region II (rectangular, including half of region IV),

$$A = (10 \text{ cm})(2.5 \text{ cm}) = 25 \text{ cm}^2$$

$$I_c = \frac{bh^3}{12} = \frac{(10 \text{ cm})(2.5 \text{ cm})^3}{12} = 13.021 \text{ cm}^4$$

$$d = 4 \text{ cm} + \frac{2.5 \text{ cm}}{2} = 5.25 \text{ cm}$$

$$d^2 A = (5.25 \text{ cm})^2 (25 \text{ cm}^2) = 689.0625 \text{ cm}^4$$

For region III (semicircular, including half of region IV),

$$A = \frac{\pi r^2}{2} = \frac{\pi (5 \text{ cm})^2}{2} = \frac{25\pi}{2}$$

$$I_c = 0.1098 r^4 = (0.1098)(5 \text{ cm})^4 = 68.625 \text{ cm}^4$$

$$d = 4 \text{ cm} + 2.5 \text{ cm} + \frac{4r}{3\pi}$$

$$= 6.5 \text{ cm} + \frac{(4)(5 \text{ cm})}{3\pi}$$

$$= 6.5 \text{ cm} + \frac{20}{3\pi} \text{ cm}$$

$$d^2 A = \left(6.5 \text{ cm} + \frac{20}{3\pi} \text{ cm}\right)^2 \left(\frac{25\pi}{2} \text{ cm}^2\right) = 2919.33 \text{ cm}^4$$

For region IV (circular),

$$A = \pi r^2 = \pi(1 \text{ cm})^2 = \pi \text{ cm}^2$$

$$I_c = \frac{\pi r^4}{4} = \frac{\pi(1 \text{ cm})^4}{4}$$

$$= \frac{\pi}{4} \text{ cm}^4$$

$$d = 4 \text{ cm} + 2.5 \text{ cm}$$

$$= 6.5 \text{ cm}$$

$$d^2 A = (6.5 \text{ cm})^2 (\pi \text{ cm}^2)$$

$$= 42.25\pi \text{ cm}^4$$

Use the parallel axis theorem for each shape.

$$I_x = \sum I_{x_c} + \sum d^2 A$$

$$= 10.67 \text{ cm}^4 + 32 \text{ cm}^4$$

$$+ 13.021 \text{ cm}^4 + 689.0625 \text{ cm}^4$$

$$+ 68.625 \text{ cm}^4 + 2919.33 \text{ cm}^4$$

$$- \frac{\pi}{4} \text{ cm}^4 - 42.25\pi \text{ cm}^4$$

$$= 3599.2 \text{ cm}^4 \quad (3600 \text{ cm}^4)$$

The answer is (C).

2. For a circle,

$$I_{x_c} = I_{y_c} = \frac{\pi r^4}{4}$$

Use the parallel axis theorem for a composite area.

$$I_{x_c} = (2)\left(\frac{\pi r^4}{4} + (\pi r^2) r^2\right) = \frac{5\pi r^4}{2}$$

$$I_{y_c} = (2)\left(\frac{\pi r^4}{4}\right) = \frac{\pi r^4}{2}$$

$$J_c = I_{x_c} + I_{y_c} = \frac{5\pi r^4}{2} + \frac{\pi r^4}{2}$$

$$= 3\pi r^4 = 3\pi(6.5 \text{ m})^4$$

$$= 16\,824 \text{ m}^4 \quad (17\,000 \text{ m}^4)$$

The answer is (D).

3. Use the formula for the moment of inertia about an edge for a rectangular shape.

For rectangle HCBA,

$$I_{x,1} = \frac{bh^3}{3} = \frac{(1.5 \text{ cm})(4.5 \text{ cm})^3}{3}$$

$$= 45.56 \text{ cm}^4$$

($I_c = bh^3/12$ could have been used, but the parallel axis theorem would also have to be used.)

Use the parallel axis theorem to calculate the moment of inertia of rectangle FEDG. $d = 6.25$ cm is the distance from the centroid of FEDG to the x-axis.

$$I_{x,2} = \frac{bh^3}{12} + Ad^2$$

$$= \frac{(7 \text{ cm})(3.5 \text{ cm})^3}{12} + (7 \text{ cm})(3.5 \text{ cm})(6.25 \text{ cm})^2$$

$$= 982 \text{ cm}^4$$

The moment of inertia for the total area is

$$I_x = I_{x,1} + I_{x,2}$$

$$= 45.56 \text{ cm}^4 + 982 \text{ cm}^4$$

$$= 1028 \text{ cm}^4 \quad (1000 \text{ cm}^4)$$

The answer is (D).

4. The length of the perimeter is

$$L = 5 \text{ cm} + 0.5 \text{ cm} + 4.5 \text{ cm}$$

$$+ 2.5 \text{ cm} + 0.5 \text{ cm} + 3 \text{ cm}$$

$$= 16 \text{ cm}$$

The x-coordinate of the centroid of the line is

$$x_c = \frac{\sum x_i L}{L}$$

$$= \left(\frac{1}{16 \text{ cm}}\right)\big((2.5 \text{ cm})(5 \text{ cm}) + (5 \text{ cm})(0.5 \text{ cm})$$

$$+ (2.75 \text{ cm})(5 \text{ cm} - 0.5 \text{ cm})$$

$$+ (0.5 \text{ cm})(3 \text{ cm} - 0.5 \text{ cm})$$

$$+ (0.25 \text{ cm})(0.5 \text{ cm}) + (0 \text{ cm})(3 \text{ cm})\big)$$

$$= 1.80 \text{ cm}$$

The answer is (D).

5. The shape is symmetrical about a vertical axis, so determine the location of the centroid by inspection. Determine the centroidal moment of inertia about the y-axis. Move the y-axis 4.1 cm to the right, so that the y-axis passes through the centroid. The centroidal y-axis passes through the centroid of the area, so the parallel axis theorem is not needed. Since the centroids of the individual rectangles coincide with the centroid of the composite area about the centroidal y-axis, I_{y_c} is simply the sum of the moments of inertia of the individual areas.

$$I_{y_c} = \frac{(4.1 \text{ cm})^3 (2.9 \text{ cm})}{12} + \frac{(8.2 \text{ cm})^3 (3.8 \text{ cm})}{12}$$

$$= 191.26 \text{ cm}^4$$

The centroidal polar moment of inertia is

$$J_c = I_{x_c} + I_{y_c}$$
$$= 142.41 \text{ cm}^4 + 191.26 \text{ cm}^4$$
$$= 333.7 \text{ cm}^4 \quad (330 \text{ cm}^4)$$

The answer is (C).

6. Separate the shape into regions. Calculate the area and locate the centroid for each region.

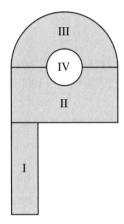

For region I (rectangular),

$$A = (3 \text{ cm})(8 \text{ cm})$$
$$= 24 \text{ cm}^2$$
$$y_c = \frac{8 \text{ cm}}{2}$$
$$= 4 \text{ cm}$$

For region II (rectangular, including half of region IV),

$$A = (2)(4 \text{ cm})(5 \text{ cm})$$
$$= 40 \text{ cm}^2$$
$$y = 8 \text{ cm} + \frac{5 \text{ cm}}{2}$$
$$= 10.5 \text{ cm}$$

For region III (semicircular, including half of region IV),

$$A = \frac{\pi r^2}{2} = \frac{\pi (4 \text{ cm})^2}{2}$$
$$= 8\pi \text{ cm}^2$$
$$y = 8 \text{ cm} + 5 \text{ cm} + \frac{4r}{3\pi}$$
$$= 13 \text{ cm} + \frac{(4)(4 \text{ cm})}{3\pi}$$
$$= 13 \text{ cm} + \frac{16}{3\pi} \text{ cm}$$

For region IV (circular),

$$A = \pi r^2 = \pi (1.5 \text{ cm})^2$$
$$= 2.25\pi \text{ cm}^2$$
$$y = 8 \text{ cm} + 5 \text{ cm} = 13 \text{ cm}$$

$$y_c = \frac{\sum y_{c,n} A_n}{\sum A_n}$$
$$= \frac{\begin{array}{c}(4 \text{ cm})(24 \text{ cm}^2) + (10.5 \text{ cm})(40 \text{ cm}^2) \\ + \left(13 \text{ cm} + \frac{16}{3\pi} \text{ cm}\right)(8\pi \text{ cm}^2) \\ - (13 \text{ cm})(2.25\pi \text{ cm}^2)\end{array}}{24 \text{ cm}^2 + 40 \text{ cm}^2 + 8\pi \text{ cm}^2 - 2.25\pi \text{ cm}^2}$$
$$= \frac{793.5 \text{ cm}^3}{82.06 \text{ cm}^2}$$
$$= 9.67 \text{ cm} \quad (9.7 \text{ cm})$$

The answer is (C).

7. The total length of the perimeter is

$$L = 2.5 \text{ cm} + 5.0 \text{ cm} + 1.0 \text{ cm} + 3.1 \text{ cm}$$
$$+ 4.5 \text{ cm} + 3.1 \text{ cm} + 1.0 \text{ cm} + 5.0 \text{ cm}$$
$$= 25.2 \text{ cm}$$

Starting with line segment FG and working counterclockwise, the x-coordinate of the centroid of the perimeter is

$$x_c = \frac{\sum x_i L_i}{L}$$
$$= \frac{\begin{array}{c}(0 \text{ cm})(3.1 \text{ cm}) + (0.5 \text{ cm})(1.0 \text{ cm}) \\ + (1.0 \text{ cm})(5.0 \text{ cm}) \\ + (2.25 \text{ cm})(2.5 \text{ cm}) \\ + (3.5 \text{ cm})(5.0 \text{ cm}) \\ + (4 \text{ cm})(1 \text{ cm}) \\ + (4.5 \text{ cm})(3.1 \text{ cm}) + (2.25 \text{ cm})(4.5 \text{ cm})\end{array}}{25.2 \text{ cm}}$$
$$= 2.25 \text{ cm} \quad (2.3 \text{ cm})$$

The y-coordinate is

$$y_c = \sum y_i L_i$$
$$= \frac{\begin{array}{c}(0 \text{ cm})(2.5 \text{ cm}) + (2)(2.5 \text{ cm})(5.0 \text{ cm}) \\ + (2)(5.0 \text{ cm})(1 \text{ cm}) + (2)(6.55 \text{ cm})(3.1 \text{ cm}) \\ + (8.1 \text{ cm})(4.5 \text{ cm})\end{array}}{25.2 \text{ cm}}$$
$$= 4.447 \text{ cm} \quad (4.4 \text{ cm})$$

The answer is (C).

CENTROIDS AND MOMENTS OF INERTIA

8. The polar radius of gyration is

$$r_p = \sqrt{\frac{J}{A}} = \sqrt{\frac{3\pi r^4}{2\pi r^2}}$$

$$= r\sqrt{\frac{3}{2}} = (2.5 \text{ m})\sqrt{\frac{3}{2}}$$

$$= 3.1 \text{ m}$$

The answer is (D).

9. Use the parallel axis theorem to find the centroidal moment of inertia of each rectangular area.

$$\text{HCBA} = \text{area 1} = (3 \text{ cm})(6 \text{ cm}) = 18 \text{ cm}^2$$
$$\text{FEDG} = \text{area 2} = (10 \text{ cm})(3 \text{ cm}) = 30 \text{ cm}^2$$

$$I_{x_c} = (I_{c,1} + A_1 d_1^2) + (I_{c,2} + A_2 d_2^2)$$

$$= \left(\frac{(3 \text{ cm})(6 \text{ cm})^3}{12} + (18 \text{ cm}^2)(5.8125 \text{ cm} - 3.0 \text{ cm})^2 \right)$$

$$+ \left(\frac{(10 \text{ cm})(3 \text{ cm})^3}{12} + (30 \text{ cm}^2)(7.5 \text{ cm} - 5.8125 \text{ cm})^2 \right)$$

$$= 304.3 \text{ cm}^4 \quad (300 \text{ cm}^4)$$

The answer is (D).

10. Separate the shape into regions. Calculate the area and locate the centroid for each region.

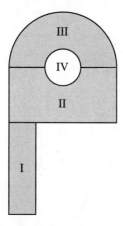

For region I (rectangular),

$$A = (1.5 \text{ cm})(5 \text{ cm})$$
$$= 7.5 \text{ cm}^2$$
$$x_c = \frac{1.5 \text{ cm}}{2}$$
$$= 0.75 \text{ cm}$$

For region II (rectangular, including half of region IV),

$$A = (3 \text{ cm})(3 \text{ cm} + 3 \text{ cm})$$
$$= 18 \text{ cm}^2$$
$$x_c = \frac{(2)(3 \text{ cm})}{2}$$
$$= 3 \text{ cm}$$

For region III (semicircular, including half of region IV),

$$A = \frac{\pi r^2}{2} = \frac{\pi (3 \text{ cm})^2}{2}$$
$$= \frac{9\pi}{2} \text{ cm}^2$$
$$x_c = 3 \text{ cm} \quad \text{[by inspection]}$$

For region IV (circular),

$$A = \pi r^2 = \pi (1 \text{ cm})^2$$
$$= \pi \text{ cm}^2$$
$$x_c = 3 \text{ cm} \quad \text{[by inspection]}$$

For the total area,

$$x_c = \frac{\sum x_{c,n} A_n}{\sum A_n}$$

$$= \frac{(0.75 \text{ cm})(7.5 \text{ cm}^2) + (3 \text{ cm})(18 \text{ cm}^2)}{7.5 \text{ cm}^2 + 18 \text{ cm}^2 + \frac{9\pi}{2} \text{ cm}^2 - \pi \text{ cm}^2}$$

$$= 2.54 \text{ cm} \quad (2.5 \text{ cm})$$

The answer is (B).

11. The vertical distance between the centroidal location of area A_1 and the composite area's centroid is

$$d_1 = \frac{h}{2} - y_c = \frac{10 \text{ cm}}{2} - 4.79 \text{ cm}$$

The moment of inertia is

$$I_{x_c} = (I_{x_c,1} + A_1 d_1^2) - I_{x'_c,2} - I_{x'_c,3}$$

$$= \left(\frac{(8 \text{ cm})(10 \text{ cm})^3}{12} + (8 \text{ cm})(10 \text{ cm})\left(\frac{10 \text{ cm}}{2} - 4.79 \text{ cm}\right)^2 \right)$$

$$- 73.94 \text{ cm}^4 - 32.47 \text{ cm}^4$$

$$= 563.8 \text{ cm}^4 \quad (560 \text{ cm}^4)$$

The answer is (D).

12. Divide the area into two rectangles, HCBA and FEDG. Their areas are

$$A_1 = (4 \text{ cm})(8 \text{ cm}) = 32 \text{ cm}^2$$
$$A_2 = (12 \text{ cm})(4 \text{ cm}) = 48 \text{ cm}^2$$

The total area is

$$A = A_1 + A_2 = 32 \text{ cm}^2 + 48 \text{ cm}^2$$
$$= 80 \text{ cm}^2$$

By inspection, the x-coordinates of the centroids of the rectangles are $x_{c,1} = x_{c,2} = 6$ cm. The x-coordinate of the centroid of the total area is

$$x_c = \frac{\sum x_{c,n} A_n}{A}$$
$$= \frac{(6 \text{ cm})(32 \text{ cm}^2) + (6 \text{ cm})(48 \text{ cm}^2)}{80 \text{ cm}^2}$$
$$= 6.0 \text{ cm}$$

By inspection, the y-coordinates of the rectangles are $y_{c,1} = 4$ cm and $y_{c,2} = 10$ cm. The y-coordinate of the centroid of the total area is

$$y_c = \frac{\sum y_{c,n} A_n}{A}$$
$$= \frac{(4 \text{ cm})(32 \text{ cm}^2) + (10 \text{ cm})(48 \text{ cm}^2)}{80 \text{ cm}^2}$$
$$= 7.6 \text{ cm}$$

The answer is (C).

13. The radius of gyration with respect to the y-axis is

$$r_y = \sqrt{\frac{I_y}{A}} = \sqrt{\frac{352 \text{ cm}^4}{36.5 \text{ cm}^2}}$$
$$= 3.1 \text{ cm}$$

The answer is (B).

26 Material Properties and Testing

PRACTICE PROBLEMS

1. A stress-strain diagram is shown.

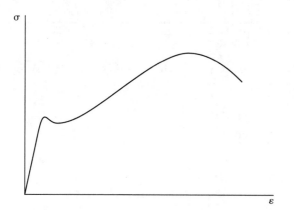

What test is represented by the illustration?

(A) resilience test
(B) rotating beam test
(C) ductility test
(D) tensile test

2. A 0.4 m long steel rod has a diameter of 0.05 m and a modulus of elasticity of 20×10^4 MPa. The rod supports a 10 000 N compressive load. What is most nearly the decrease in the steel rod's length?

(A) 1.3×10^{-6} m
(B) 2.5×10^{-6} m
(C) 5.1×10^{-6} m
(D) 1.0×10^{-5} m

3. What is the term for the ratio of stress to strain below the proportional limit?

(A) modulus of rigidity
(B) Hooke's constant
(C) Poisson's ratio
(D) Young's modulus

4. What do impact tests determine?

(A) hardness
(B) yield strength
(C) toughness
(D) creep strength

5. The density of a particular metal is 2750 kg/m^3. The modulus of elasticity for this metal is 210 GPa. A circular bar of this metal 3.5 m long and 160 cm^2 in cross-sectional area is suspended vertically from one end. What is most nearly the elongation of the bar due to its own mass?

(A) 0.00055 mm
(B) 0.00079 mm
(C) 0.0016 mm
(D) 0.0024 mm

6. A stress-strain diagram is shown.

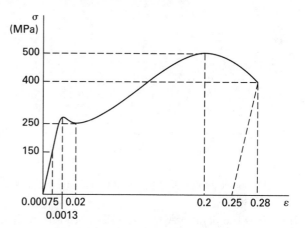

What is most nearly the percent elongation at failure?

(A) 14%
(B) 19%
(C) 25%
(D) 28%

7. What does the value of 40 MPa in the illustration shown represent?

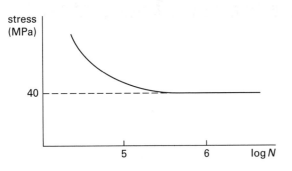

I. fatigue limit
II. endurance limit
III. proportional limit
IV. yield stress

(A) I only
(B) I and II
(C) II and IV
(D) I, II, and IV

8. If δ is deformation, and L is the original length of the specimen, what is the definition of normal strain, ε?

(A) $\varepsilon = \dfrac{L+\delta}{L}$
(B) $\varepsilon = \dfrac{L+\delta}{\delta}$
(C) $\varepsilon = \dfrac{\delta}{L+\delta}$
(D) $\varepsilon = \dfrac{\delta}{L}$

9. Which of the following statements regarding the ductile-to-brittle transition temperature is true?

I. It is important for structures used in cold environments.
II. It is the point at which the size of the shear lip or tearing rim goes to zero.
III. It is the temperature at which 20 J of energy causes failure in a Charpy V-notch specimen of standard dimensions.

(A) I only
(B) I and II
(C) I and III
(D) II and III

10. A stress-strain diagram is shown.

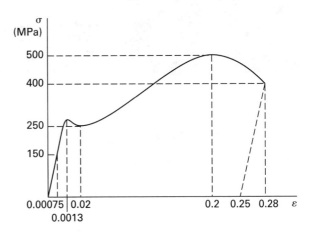

What is most nearly the modulus of elasticity of the material?

(A) 20 GPa
(B) 80 GPa
(C) 100 GPa
(D) 200 GPa

SOLUTIONS

1. The illustration shows results from a tensile test. Both resilience and ductility may be calculated from the results, but the test is not known by those names. The rotating beam is a cyclic test and does not yield a monotonic stress-strain curve.

The answer is (D).

2. The area of the steel rod is

$$A_0 = \frac{\pi d^2}{4} = \frac{\pi (0.05 \text{ m})^2}{4} = 1.96 \times 10^{-3} \text{ m}^2$$

The decrease in the rod's length is

$$\Delta L = \frac{FL_0}{A_0 E}$$

$$= \frac{(10\,000 \text{ N})(0.4 \text{ m})}{(1.96 \times 10^{-3} \text{ m}^2)(20 \times 10^4 \text{ MPa})\left(10^6 \frac{\text{Pa}}{\text{MPa}}\right)}$$

$$= 1.019 \times 10^{-5} \text{ m} \quad (1.0 \times 10^{-5} \text{ m})$$

The answer is (D).

3. Young's modulus is defined by Hooke's law.

$$\sigma = E\varepsilon$$

E is Young's modulus, or the modulus of elasticity, equal to the stress divided by strain within the proportional region of the stress-strain curve.

The answer is (D).

4. An impact test measures the energy needed to fracture the test sample. This is a toughness parameter.

The answer is (C).

5. The mass of the bar is

$$m = \rho V = \rho A L$$

$$= \left(2750 \frac{\text{kg}}{\text{m}^3}\right)\left(\frac{160 \text{ cm}^2}{\left(100 \frac{\text{cm}}{\text{m}}\right)^2}\right)(3.5 \text{ m})$$

$$= 154 \text{ kg}$$

The total gravitational force is experienced by the metal at the suspension point. Farther down the rod, however, there is less volume contributing to the force, and the stress is reduced. The average force on the metal in the bar is half of the maximum value.

$$F_{\text{ave}} = \tfrac{1}{2} F_{\text{max}} = \tfrac{1}{2} mg$$

$$= \left(\tfrac{1}{2}\right)(154 \text{ kg})\left(9.81 \frac{\text{m}}{\text{s}^2}\right)$$

$$= 755 \text{ N}$$

The elongation is

$$\varepsilon = \frac{\Delta L}{L_0}$$

$$\Delta L = \varepsilon L_0 = \frac{\sigma}{E} L_0 = \frac{F}{AE} L_0$$

$$= \left(\frac{755 \text{ N}}{(160 \text{ cm}^2)(210 \times 10^9 \text{ Pa})}\right)(3.5 \text{ m})\left(100 \frac{\text{cm}}{\text{m}}\right)^2$$

$$= 7.868 \times 10^{-7} \text{ m} \quad (0.00079 \text{ mm})$$

The answer is (B).

6. The strain at failure used in the equation is found by extending a line from the failure point to the strain axis, parallel to the linear portion of the curve. The percent elongation is an indicator of the ductility of a material, but it is not the same as the ductility. The percent elongation is

$$\text{percent elongation} = \varepsilon_f \times 100\%$$

$$= 0.25 \times 100\%$$

$$= 25\%$$

The answer is (C).

7. The illustration shows results of an endurance (or fatigue) test. The value of 40 MPa is called the endurance stress, endurance limit, or fatigue limit, and is equal to the maximum stress that can be repeated indefinitely without causing the specimen to fail.

The answer is (B).

8. Strain is defined as elongation, δ, per unit length, L.

The answer is (D).

9. Option II is the only one that is false. A test piece that breaks at 20 J of energy usually has a small shear lip.

The answer is (C).

10. The modulus of elasticity (Young's modulus) is the slope of the stress-strain line in the proportional region.

$$\sigma = E\varepsilon$$

$$E = \frac{\sigma}{\varepsilon} = \frac{150 \text{ MPa}}{0.00075}$$

$$= 200\,000 \text{ MPa} \quad (20 \times 10^4 \text{ MPa})$$

The answer is (D).

27 Engineering Materials

PRACTICE PROBLEMS

1. Refer to the phase diagram shown.

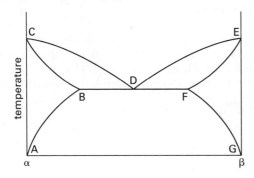

The region enclosed by points DEF can be described as a

(A) mixture of solid β component and liquid α component

(B) mixture of solid β and liquid β component

(C) peritectic composition

(D) mixture of solid β component and a molten mixture of α and β components

2. Which of the following figures is a cooling curve of a pure metal?

(A)

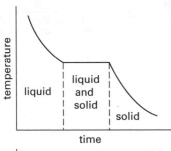

(B)

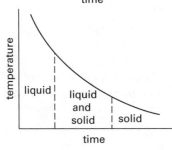

(C)

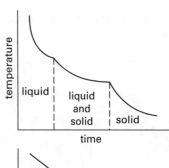

(D)

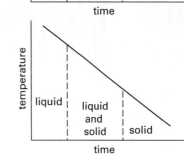

3. A composite material consists of 20 kg of material A, 10 kg of material B, and 5 kg of material C. The densities of materials A, B, and C are 2 g/cm³, 3 g/cm³, and 4 g/cm³, respectively. What is most nearly the density of the composite material?

(A) 2.1 g/cm³

(B) 2.4 g/cm³

(C) 2.7 g/cm³

(D) 3.3 g/cm³

4. Which of the following characteristics describes martensite?

I. high ductility

II. formed by quenching austenite

III. high hardness

(A) I only

(B) I and II

(C) I and III

(D) II and III

5. A mixture of ice and water is held at a constant temperature of 0°C. How many degrees of freedom does the mixture have?

(A) −1
(B) 0
(C) 1
(D) 2

6. Given the electrochemical cell shown, what is the reaction at the anode?

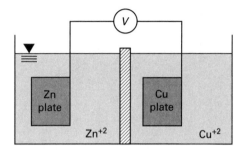

(A) $Cu \rightarrow Cu^{2+} + 2e^-$
(B) $Cu^{2+} + 2e^- \rightarrow Cu$
(C) $Zn \rightarrow Zn^{2+} + 2e^-$
(D) $Zn^{2+} + 2e^- \rightarrow Zn$

7. Refer to the phase diagram shown.

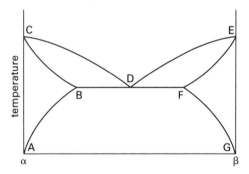

Which line(s) is/are the liquidus line(s)?

(A) CBDFG
(B) CDE
(C) CBFE
(D) ABC and EFG

8. What is the hardest form of steel?

(A) pearlite
(B) ferrite
(C) bainite
(D) martensite

9. Which of the following processes can increase the deformation resistance of steel?

I.   tempering
II. hot working
III. adding alloying elements
IV. hardening

(A) I and II
(B) I and IV
(C) II and III
(D) III and IV

10. Corrosion of iron can be inhibited with a more electropositive coating, while a less electropositive coating tends to accelerate corrosion. Which of the following coatings will contribute to corrosion of iron products?

(A) zinc
(B) gold
(C) aluminum
(D) magnesium

11. Refer to the phase diagram shown.

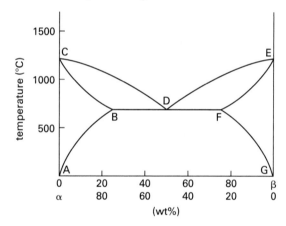

Approximately how much solid (as a percentage by weight) exists when the mixture is 30% α and 70% β and the temperature is 800°C?

(A) 0%
(B) 19%
(C) 30%
(D) 50%

12. What metal can be used as a sacrificial anode on a small ocean fishing boat with an aluminum hull?

(A) zinc
(B) magnesium
(C) iron
(D) copper

13. Galvanic corrosion between a steel pipe and an attached copper fitting is to be counteracted electrically with a monitored direct current power supply. Most nearly, what should be the applied voltage to eliminate the corrosion?

(A) 0.10 V

(B) 0.34 V

(C) 0.44 V

(D) 0.78 V

14. Which of the following methods is the most effective in reducing galvanic corrosion between faying objects?

(A) manufacturing both parts from the same material

(B) eliminating moisture in the atmosphere

(C) plating or painting one or both parts with epoxy primer

(D) lubricating the parts

SOLUTIONS

1. The region describes a mixture of solid β component and a liquid of components α and β.

The answer is (D).

2. The solidification of a molten metal is no different than the solidification of water into ice. During the phase change, the temperature remains constant as the heat of fusion is removed. The temperature remains constant during the phase change.

The answer is (A).

3. Calculate the volume, V, of each material.

$$V_A = \frac{m_A}{\rho_A} = \frac{(20 \text{ kg})\left(1000 \frac{\text{g}}{\text{kg}}\right)}{2 \frac{\text{g}}{\text{cm}^3}}$$
$$= 10\,000 \text{ cm}^3$$

$$V_B = \frac{m_B}{\rho_B} = \frac{(10 \text{ kg})\left(1000 \frac{\text{g}}{\text{kg}}\right)}{3 \frac{\text{g}}{\text{cm}^3}}$$
$$= 3333 \text{ cm}^3$$

$$V_C = \frac{m_C}{\rho_C} = \frac{(5 \text{ kg})\left(1000 \frac{\text{g}}{\text{kg}}\right)}{4 \frac{\text{g}}{\text{cm}^3}}$$
$$= 1250 \text{ cm}^3$$

The total volume is

$$V_{tot} = V_A + V_B + V_C$$
$$= 10\,000 \text{ cm}^3 + 3333 \text{ cm}^3 + 1250 \text{ cm}^3$$
$$= 14\,583 \text{ cm}^3$$

The density of the composite material is

$$\rho_c = \sum f_i \rho_i = f_A \rho_A + f_B \rho_B + f_C \rho_C$$
$$= \frac{V_A \rho_A + V_B \rho_B + V_C \rho_C}{V_{tot}}$$
$$= \frac{(10\,000 \text{ cm}^3)\left(2 \frac{\text{g}}{\text{cm}^3}\right) + (3333 \text{ cm}^3)\left(3 \frac{\text{g}}{\text{cm}^3}\right) + (1250 \text{ cm}^3)\left(4 \frac{\text{g}}{\text{cm}^3}\right)}{14\,583 \text{ cm}^3}$$
$$= 2.4 \text{ g/cm}^3$$

The answer is (B).

4. Martensite is a hard, strong, and brittle material formed by rapid cooling of austenite.

The answer is (D).

5. Since solid and liquid phases are present simultaneously, the number of phases, P, is 2. Only water is involved, so the number of compounds, C, is 1.

Gibbs' phase rule is applicable when both temperature and pressure can be varied. When the temperature is held constant, Gibbs' phase rule is

$$P + F = C + 1\big|_{\text{constant temperature}}$$
$$F = C + 1 - P$$
$$= 1 + 1 - 2$$
$$= 0$$

The answer is (B).

6. Zinc has a higher potential and will act as the anode. By definition, the anode is where electrons are lost. The reaction at the anode of the electrochemical cell is $Zn \rightarrow Zn^{2+} + 2e^-$.

The answer is (C).

7. The liquidus line divides the diagram into two regions. Above the liquidus line, the alloy is purely liquid, while below the liquidus line, the alloy may exist as solid phase or as a mixture of solid and liquid phases. The liquidus line is CDE.

The answer is (B).

8. Hard steel is obtained by rapid quenching. Martensite has a high hardness since it is rapidly quenched. Though martensite is hard, it has low ductility.

The answer is (D).

9. Surface hardening processes will increase the deformation resistance of steel. Some alloying metals will also increase steel hardness. Tempering and hot working increase the ductility (deformation capability) of steel.

The answer is (D).

10. Zinc, aluminum, and magnesium are all more electropositive (anodic) than iron and will corrode sacrificially to protect it. Gold is more cathodic and will be protected at the expense of the iron.

The answer is (B).

11. Use the phase diagram to find the fraction of solid.

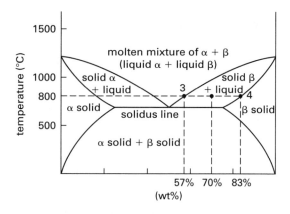

$$\text{wt\% fraction solid} = \frac{x - x_3}{x_4 - x_3} \times 100\%$$
$$= \frac{70\% - 57\%}{83\% - 57\%} \times 100\%$$
$$= 50\%$$

The answer is (D).

12. Aluminum is near the bottom of the galvanic scale. Only magnesium is lower, more anodic, and has a greater anode half-cell potential. Therefore, magnesium will become the anode, while the aluminum will become the cathode. The less noble magnesium gives up its electrons to the more noble aluminum. During this process, the magnesium breaks down (corrodes), resulting in magnesium oxide plus free electrons flowing through the electrolyte (sea water) to the aluminum.

The answer is (B).

13. The oxidation potential of the iron in the steel pipe is 0.440 V; the oxidation potential of the copper is -0.337 V. The difference in potentials is 0.440 V $-$ (-0.337 V) $= 0.777$ V (0.78 V).

The answer is (D).

14. Faying parts rub against each other. Any paint, plating, coating, or lubricant will be abraded off. Since the parts are already in contact, an electrolyte (moisture) is not needed to complete the circuit. Galvanic action will be reduced if the parts are manufactured from the same material.

The answer is (A).

28 Manufacturing Processes

PRACTICE PROBLEMS

1. A high-speed steel tool has a 5 h life when operated at 1.5 m/s. What is most nearly the tool life if the speed is reduced to 1.4 m/s?

(A) 2.5 h
(B) 6.6 h
(C) 7.1 h
(D) 10 h

2. A joint is made between two metal pieces by closely fitting their surfaces and distributing a molten nonferrous filler metal to the interface by capillary attraction. The pieces to be joined have a melting point of 750°C, and the filler metal melts at 375°C. This process is most accurately termed

(A) soldering
(B) brazing
(C) welding
(D) forge welding

3. Which procedure involves molten metal?

(A) rolling
(B) coining
(C) piercing
(D) casting

4. Which of the following grinding procedures results in essentially NO change in the dimensions of the workpiece?

(A) superfinishing
(B) lapping
(C) honing
(D) all of the above

5. Which operation is NOT a cold-working operation?

(A) milling
(B) grinding
(C) drilling
(D) piercing

6. Which type of welding uses coated electrodes?

(A) SMAW
(B) TIG
(C) MIG
(D) GMAW

SOLUTIONS

1. $n = 0.1$ for high-speed tool steel. Rearranging Taylor's equation,

$$v_1 T_1^n = v_2 T_2^n$$

$$T_2 = T_1 \left(\frac{v_1}{v_2}\right)^{1/n} = (5 \text{ h}) \left(\frac{1.5 \frac{\text{m}}{\text{s}}}{1.4 \frac{\text{m}}{\text{s}}}\right)^{1/0.1}$$

$$= 9.968 \text{ h} \quad (10 \text{ h})$$

The answer is (D).

2. Definitions from the American Welding Society are:

Soldering—A joining process wherein coalescence between metal parts is produced by heating to suitable temperatures generally below 425°C and by using nonferrous filler metals having melting temperatures below those of the base metals. The solder is usually distributed between the properly fitted surfaces of the joint by capillary attraction.

Brazing—A group of welding processes wherein coalescence is produced by heating to suitable temperatures above 425°C and by using a nonferrous filler metal having a melting point below those of the base metals. The filler metal is distributed between closely fitted surfaces of the joint by capillary action.

Welding—A localized coalescence of metals wherein coalescence is produced by heating to suitable temperatures, with or without the application of pressure, and with or without the use of filler metal. The filler metal either has a melting point approximately the same as the base metals, or has a melting point below that of the base metals but above 425°C.

Forge welding—A group of welding processes wherein coalescence is produced by heating in a forge or other furnace and by applying pressure or blows. An example is the hammer welding process previously used in railroad and blacksmith shops.

The answer is (A).

3. In casting, molten metal is forced into a mold.

The answer is (D).

4. Superfinishing or ultrafinishing is used to remove the smear metal left on a piece after lapping, honing, or other grinding procedures have been performed. The change in material dimensions caused by superfinishing is negligible.

The answer is (A).

5. Milling, grinding, and drilling are almost always performed at ambient temperatures, much below the recrystallization temperature of a metal. There are two definitions of piercing, a cold-working process which is essentially the same as blanking or die-cutting, and a hot-working process in which a heated steel billet is pierced longitudinally to produce seamless pipe.

The answer is (D).

6. Shielded metal arc welding (SMAW) uses electrodes coated with SiO_2 and/or TiO_2, as well as other metal oxides.

The answer is (A).

29 Stresses and Strains

PRACTICE PROBLEMS

1. The element is subjected to the plane stress condition shown.

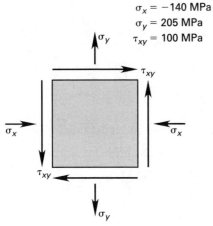

What is the maximum shear stress?

(A) 100 MPa
(B) 160 MPa
(C) 200 MPa
(D) 210 MPa

2. A plane element in a body is subjected to a normal tensile stress in the x-direction of 84 MPa, as well as shear stresses of 28 MPa, as shown.

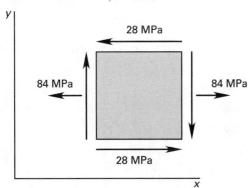

Most nearly, what are the principal stresses?

(A) 70 MPa; 14 MPa
(B) 84 MPa; 28 MPa
(C) 92 MPa; −8.5 MPa
(D) 112 MPa; −28 MPa

3. What is most nearly the lateral strain, ε_y, of the steel specimen shown if $F_x = 3000$ kN, $E = 193$ GPa, and $\nu = 0.29$?

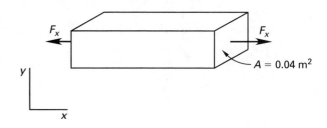

(A) -4.0×10^{-4}
(B) -1.1×10^{-4}
(C) 1.0×10^{-4}
(D) 4.0×10^{-4}

4. The elements are subjected to the plane stress condition shown. The maximum shear stress is 109.2 MPa.

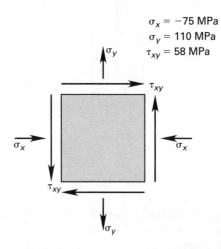

What are the orientations of the stress planes (relative to the x-axis)?

(A) −74°; 15°
(B) −58°; 32°
(C) −32°; 58°
(D) −16°; 74°

5. What is most nearly the elongation of the aluminum bar (cross section of 3 cm × 3 cm) shown when loaded to its yield point? The modulus of elasticity is 69 GPa, and the yield strength in tension is 255 MPa. Neglect the weight of the bar.

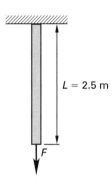

(A) 3.3 mm
(B) 9.3 mm
(C) 12 mm
(D) 15 mm

6. The column shown has a cross-sectional area of 13 m².

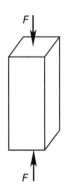

What is the approximate maximum load if the compressive stress cannot exceed 9.6 kPa?

(A) 120 kN
(B) 122 kN
(C) 125 kN
(D) 130 kN

7. The element is subjected to the plane stress condition shown. The maximum shear stress is 300 MPa.

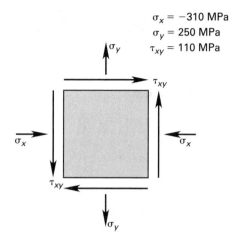

The principal stresses are most nearly

(A) 250 MPa; −310 MPa
(B) 270 MPa; −330 MPa
(C) 330 MPa; −270 MPa
(D) 310 MPa; −250 MPa

8. Given a shear stress of $\tau_{xy} = 35$ MPa and a shear modulus of $G = 75$ GPa, the shear strain is most nearly

(A) 2.5×10^{-5} rad
(B) 4.7×10^{-4} rad
(C) 5.5×10^{-4} rad
(D) 8.3×10^{-4} rad

9. Which of the following could be the Poisson ratio of a material?

(A) 0.35
(B) 0.52
(C) 0.55
(D) 0.60

10. A plane element in a body is subjected to the stresses shown.

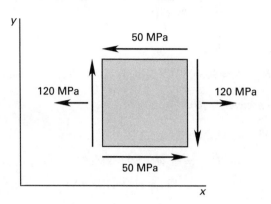

What is most nearly the maximum shear stress?

(A) 50 MPa
(B) 64 MPa
(C) 72 MPa
(D) 78 MPa

SOLUTIONS

1. There are two methods for solving the problem. The first method is to use the equation for τ_{max}; the second method is to draw Mohr's circle.

Solving by the equation for τ_{max},

$$\tau_{max} = \pm\sqrt{\left(\frac{\sigma_x - \sigma_y}{2}\right)^2 + \tau_{xy}^2}$$

$$= \sqrt{\left(\frac{-140 \text{ MPa} - 205 \text{ MPa}}{2}\right)^2 + (100 \text{ MPa})^2}$$

$$= 199.4 \text{ MPa} \quad (200 \text{ MPa})$$

Solving by Mohr's circle,

step 1:
$$\sigma_x = -140 \text{ MPa}$$
$$\sigma_y = 205 \text{ MPa}$$
$$\tau_{xy} = 100 \text{ MPa}$$

step 2: Draw σ-τ axes.

step 3: The circle center is

$$C = \tfrac{1}{2}(\sigma_x + \sigma_y)$$
$$= \left(\tfrac{1}{2}\right)(-140 \text{ MPa} + 205 \text{ MPa})$$
$$= 32.5 \text{ MPa}$$

step 4: Plot the points $(-140 \text{ MPa}, -100 \text{ MPa})$ and $(205 \text{ MPa}, 100 \text{ MPa})$.

step 5: Draw the diameter of the circle.

step 6: Draw the circle.

step 7: Find the radius of the circle.

step 8: Maximum shear stress is at the top of the circle, $\tau_{max} = 199.4$ MPa (200 MPa).

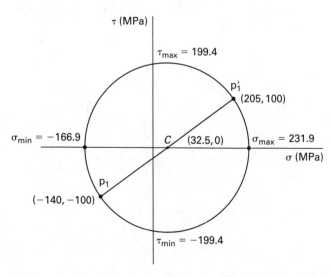

The answer is (C).

2. τ_{xy} is negative according to the standard sign convention.

$$\sigma_{max,min} = \tfrac{1}{2}(\sigma_x + \sigma_y) \pm \sqrt{\left(\frac{\sigma_x - \sigma_y}{2}\right)^2 + \tau_{xy}^2}$$

$$= \left(\tfrac{1}{2}\right)(84 \text{ MPa} + 0 \text{ MPa})$$

$$\pm \sqrt{\left(\frac{84 \text{ MPa} - 0 \text{ MPa}}{2}\right)^2 + (-28 \text{ MPa})^2}$$

$$= 42 \text{ MPa} \pm 50.478 \text{ MPa}$$

$$= 92.478 \text{ MPa};\ -8.478 \text{ MPa}$$

$$(92 \text{ MPa};\ -8.5 \text{ MPa})$$

The answer is (C).

3. From Hooke's law and the equation for axial stress,

$$\varepsilon_x = \frac{\sigma_x}{E} = \frac{F_x}{EA} = \frac{(3000 \text{ kN})\left(1000 \frac{\text{N}}{\text{kN}}\right)}{(193 \text{ GPa})\left(10^9 \frac{\text{Pa}}{\text{GPa}}\right)(0.04 \text{ m}^2)}$$

$$= 3.89 \times 10^{-4}$$

Use Poisson's ratio.

$$\varepsilon_y = -\nu\varepsilon_x = (-0.29)(3.89 \times 10^{-4})$$

$$= -1.13 \times 10^{-4} \quad (-1.1 \times 10^{-4})$$

The answer is (B).

4. Calculate the angles, θ, of the stress planes.

$$\theta = \tfrac{1}{2}\arctan\frac{2\tau_{xy}}{\sigma_x - \sigma_y}$$

$$= \tfrac{1}{2}\arctan\frac{(2)(58 \text{ MPa})}{-75 \text{ MPa} - 110 \text{ MPa}}$$

$$= -16.04°;\ 73.96° \quad (-16°;\ 74°)$$

Alternatively, the orientations can be found graphically from Mohr's circle.

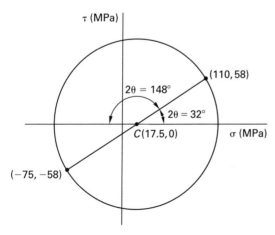

The answer is (D).

5. From Hooke's law, the axial strain is

$$\varepsilon = \frac{\sigma}{E} = \frac{(255 \text{ MPa})\left(10^6 \frac{\text{Pa}}{\text{MPa}}\right)}{(69 \text{ GPa})\left(10^9 \frac{\text{Pa}}{\text{GPa}}\right)} = 0.0037$$

The elongation is

$$\delta = \varepsilon L = (0.0037)(2.5 \text{ m}) = 0.00925 \text{ m} \quad (9.3 \text{ mm})$$

The answer is (B).

6. The maximum force is

$$F_{max} = S_a A = (9.6 \text{ kPa})(13 \text{ m}^2)$$

$$= 124.8 \text{ kN} \quad (125 \text{ kN})$$

The answer is (C).

7. The principal stresses are

$$\sigma_{max}, \sigma_{min} = \tfrac{1}{2}(\sigma_x + \sigma_y) \pm \tau_{max}$$

$$= \left(\tfrac{1}{2}\right)(-310 \text{ MPa} + 250 \text{ MPa}) \pm 300 \text{ MPa}$$

$$= -30 \text{ MPa} \pm 300 \text{ MPa}$$

$$\sigma_{max} = 270 \text{ MPa}$$

$$\sigma_{min} = -330 \text{ MPa}$$

The answer is (B).

8. Use Hooke's law for shear.

$$\gamma = \frac{\tau_{xy}}{G} = \frac{(35 \text{ MPa})\left(10^6 \frac{\text{Pa}}{\text{MPa}}\right)}{(75 \text{ GPa})\left(10^9 \frac{\text{Pa}}{\text{GPa}}\right)}$$

$$= 4.67 \times 10^{-4} \text{ rad} \quad (4.7 \times 10^{-4} \text{ rad})$$

The answer is (B).

9. The Poisson ratio is almost always in the range $0 < \nu < 0.5$. Option A (0.35) is the only answer that satisfies this condition.

The answer is (A).

10. The maximum shear stress is

$$\tau_{max} = \sqrt{\left(\frac{\sigma_x - \sigma_y}{2}\right)^2 + \tau_{xy}^2}$$

$$= \sqrt{\left(\frac{120 \text{ MPa} - 0 \text{ MPa}}{2}\right)^2 + (-50 \text{ MPa})^2}$$

$$= 78.10 \text{ MPa} \quad (78 \text{ MPa})$$

The answer is (D).

30 Thermal, Hoop, and Torsional Stress

PRACTICE PROBLEMS

1. The maximum torque on a 0.15 m diameter solid shaft is 13 500 N·m. What is most nearly the maximum shear stress in the shaft?

(A) 20 MPa

(B) 23 MPa

(C) 28 MPa

(D) 34 MPa

2. The unrestrained glass window shown is subjected to a temperature change from 0°C to 50°C. The coefficient of thermal expansion for the glass is 8.8×10^{-6} 1/°C.

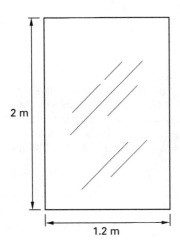

What is most nearly the change in area of the glass?

(A) 0.00040 m²

(B) 0.0013 m²

(C) 0.0021 m²

(D) 0.0028 m²

3. The cylindrical steel tank shown is 3.5 m in diameter, 5 m high, and filled to the top with a brine solution. Brine has a density of 1198 kg/m³. The thickness of the steel shell is 12.5 mm. Neglect the weight of the tank.

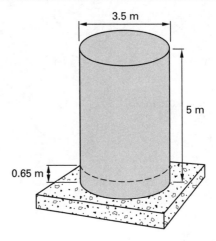

What is the approximate hoop stress in the steel 0.65 m above the rigid concrete pad?

(A) 1.2 MPa

(B) 1.4 MPa

(C) 7.2 MPa

(D) 11 MPa

4. A steel shaft is shown. The shear modulus is 80 GPa.

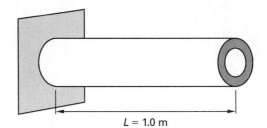

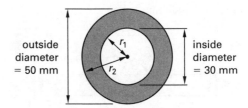

Most nearly, what torque should be applied to the end of the shaft in order to produce a twist of 1.5°?

(A) 420 N·m

(B) 560 N·m

(C) 830 N·m

(D) 1100 N·m

5. For the shaft shown, the shear stress is not to exceed 110 MPa.

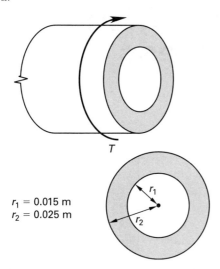

$r_1 = 0.015$ m
$r_2 = 0.025$ m

What is most nearly the largest torque that can be applied?

(A) 1700 N·m
(B) 1900 N·m
(C) 2300 N·m
(D) 3400 N·m

6. An aluminum (shear modulus $= 2.8 \times 10^{10}$ Pa) rod is 25 mm in diameter and 50 cm long. One end is rigidly fixed to a support. Most nearly, what torque must be applied at the free end to twist the rod 4.5° about its longitudinal axis?

(A) 26 N·m
(B) 84 N·m
(C) 110 N·m
(D) 170 N·m

7. A circular bar at 10°C is constrained by rigid concrete walls at both ends. The bar is 1000 mm long and has a cross-sectional area of 2600 mm².

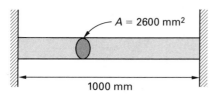

E = modulus of elasticity
 $= 200$ GPa
α = coefficient of thermal expansion
 $= 9.4 \times 10^{-6}$ 1/°C

What is most nearly the axial force in the bar if the temperature is raised to 40°C?

(A) 92 kN
(B) 110 kN
(C) 130 kN
(D) 150 kN

8. A 3 m diameter solid bar experiences opposing torques of 280 N·m at each end.

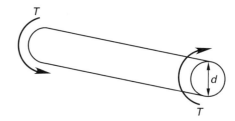

What is most nearly the maximum shear stress in the bar?

(A) 2.2 Pa
(B) 31 Pa
(C) 42 Pa
(D) 53 Pa

9. A 12.5 mm diameter steel rod is pinned between two rigid walls. The rod is initially unstressed. The rod's temperature subsequently increases 50°C. The rod is adequately stiffened and supported such that buckling does not occur. The coefficient of linear thermal expansion for steel is 11.7×10^{-6} 1/°C. The modulus of elasticity for steel is 210 GPa.

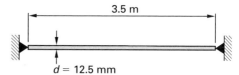

What is the approximate axial force in the rod?

(A) 2.8 kN
(B) 15 kN
(C) 19 kN
(D) 58 kN

10. 10 km of steel railroad track are placed when the temperature is 20°C. The linear coefficient of thermal expansion for the rails is 11×10^{-6} 1/°C. The track is free to slide forward. Most nearly, how far apart will the ends of the track be when the temperature reaches 50°C?

(A) 10.0009 km
(B) 10.0027 km
(C) 10.0033 km
(D) 10.0118 km

11. A deep-submersible diving bell has a cylindrical pressure hull with an outside diameter of 3.5 m and a wall thickness of 15 cm constructed from a ductile material. The hull is expected to experience an external pressure of 50 MPa. The hull should be designed as a

(A) thin-walled pressure vessel using the outer radius in the stress calculations
(B) thin-walled pressure vessel using the logarithmic mean area in stress calculations
(C) thin-walled pressure vessel using factors of safety of at least 4 for ductile materials and at least 8 for brittle components such as viewing ports
(D) thick-walled pressure vessel

12. A cantilever horizontal hollow tube is acted upon by a vertical force and a torque at its free end.

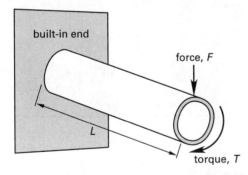

Where is the maximum stress in the cylinder?

(A) at the upper surface at midlength ($L/2$)
(B) at the lower surface at the built-in end
(C) at the upper surface at the built-in end
(D) at both the upper and lower surfaces at the built-in end

13. One end of the hollow aluminum shaft is fixed, and the other end is connected to a gear with an outside diameter of 40 cm as shown. The gear is subjected to a tangential gear force of 45 kN. The shear modulus of the aluminum is 2.8×10^{10} Pa.

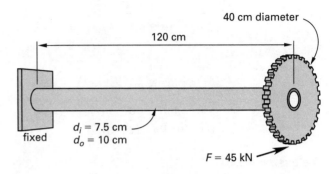

What are most nearly the maximum angle of twist and the shear stress in the shaft?

(A) 0.016 rad, 14 MPa
(B) 0.025 rad, 220 MPa
(C) 0.057 rad, 67 MPa
(D) 0.25 rad, 200 MPa

14. A compressed gas cylinder for use in a laboratory has an internal gage pressure of 8 MPa at the time of delivery. The outside diameter of the cylinder is 25 cm. If the steel has an allowable stress of 90 MPa, what is the required thickness of the wall?

(A) 0.69 cm
(B) 0.95 cm
(C) 1.1 cm
(D) 1.9 cm

SOLUTIONS

1. The polar moment of inertia is

$$J = \frac{\pi r^4}{2} = \left(\frac{\pi}{2}\right)\left(\frac{0.15 \text{ m}}{2}\right)^4$$
$$= 4.97 \times 10^{-5} \text{ m}^4$$

The shear stress is

$$\tau = \frac{Tr}{J} = \frac{(13\,500 \text{ N·m})\left(\frac{0.15 \text{ m}}{2}\right)}{4.97 \times 10^{-5} \text{ m}^4}$$
$$= 20.37 \times 10^6 \text{ Pa} \quad (20 \text{ MPa})$$

The answer is (A).

2. Changes in temperature affect each linear dimension.

$$\delta_{\text{width}} = \alpha L(T - T_o)$$
$$= \left(8.8 \times 10^{-6} \, \frac{1}{°\text{C}}\right)(1.2 \text{ m})(50°\text{C} - 0°\text{C})$$
$$= 0.000528 \text{ m}$$
$$\delta_{\text{height}} = \left(8.8 \times 10^{-6} \, \frac{1}{°\text{C}}\right)(2 \text{ m})(50°\text{C} - 0°\text{C})$$
$$= 0.00088 \text{ m}$$

$$A_{\text{initial}} = (2 \text{ m})(1.2 \text{ m}) = 2.4 \text{ m}^2$$
$$A_{\text{final}} = (2 \text{ m} + 0.00088 \text{ m})$$
$$\quad \times (1.2 \text{ m} + 0.000528 \text{ m})$$
$$= 2.40211 \text{ m}^2$$
$$\Delta A = A_{\text{final}} - A_{\text{initial}}$$
$$= 2.40211 \text{ m}^2 - 2.4 \text{ m}^2$$
$$= 0.00211 \text{ m}^2 \quad (0.0021 \text{ m}^2)$$

Alternative Solution

The area coefficient of thermal expansion is, for all practical purposes, equal to 2α.

The change in area is

$$\Delta A = 2\alpha A_o \Delta T$$
$$= (2)\left(8.8 \times 10^{-6} \, \frac{1}{°\text{C}}\right)(2.4 \text{ m}^2)(50°\text{C} - 0°\text{C})$$
$$= 0.00211 \text{ m}^2 \quad (0.0021 \text{ m}^2)$$

The answer is (C).

3. Determine whether the tank is thin-walled or thick-walled.

$$\frac{t}{r} = \frac{12.5 \text{ mm}}{\left(\frac{3.5 \text{ m}}{2}\right)\left(1000 \, \frac{\text{mm}}{\text{m}}\right)} = 0.007 < 0.1$$

Use formulas for thin-walled cylindrical tanks. The pressure is

$$p = \rho g h$$
$$= \left(1198 \, \frac{\text{kg}}{\text{m}^3}\right)\left(9.81 \, \frac{\text{m}}{\text{s}^2}\right)(5 \text{ m} - 0.65 \text{ m})$$
$$= 51\,123 \text{ Pa}$$

The hoop stress is

$$\sigma_t = \frac{pr}{t} = \frac{pd}{2t} = \frac{(51\,123 \text{ Pa})(3.5 \text{ m})}{(2)\left(\frac{12.5 \text{ mm}}{1000 \, \frac{\text{mm}}{\text{m}}}\right)}$$
$$= 7.157 \times 10^6 \text{ Pa} \quad (7.2 \text{ MPa})$$

The answer is (C).

4. Convert the twist angle to radians.

$$\phi = (1.5°)\left(\frac{2\pi \text{ rad}}{360°}\right) = 0.026 \text{ rad}$$

Calculate the polar moment of inertia, J.

$$r_1 = 15 \text{ mm} \quad (0.015 \text{ m})$$
$$r_2 = 25 \text{ mm} \quad (0.025 \text{ m})$$
$$J = \frac{\pi}{2}(r_o^4 - r_i^4) = \left(\frac{\pi}{2}\right)\left((0.025 \text{ m})^4 - (0.015 \text{ m})^4\right)$$
$$= 5.34 \times 10^{-7} \text{ m}^4$$

The torque is

$$T = \frac{\phi G J}{L}$$
$$= \frac{(0.026 \text{ rad})(80 \text{ GPa})\left(10^9 \, \frac{\text{Pa}}{\text{GPa}}\right)}{1.0 \text{ m}}$$
$$\quad \times (5.34 \times 10^{-7} \text{ m}^4)$$
$$= 1119 \text{ N·m} \quad (1100 \text{ N·m})$$

The answer is (D).

5. Since the shear stress is largest at the outer diameter, the maximum torque is found using this radius. For an annular region,

$$J = \frac{\pi}{2}(r_o^4 - r_i^4) = \left(\frac{\pi}{2}\right)\left((0.025 \text{ m})^4 - (0.015 \text{ m})^4\right)$$
$$= 5.34 \times 10^{-7} \text{ m}^4$$

THERMAL, HOOP, AND TORSIONAL STRESS 30-5

The torque is

$$T_{\max} = \frac{\tau J}{r_2} = \frac{(110 \text{ MPa})\left(10^6 \frac{\text{Pa}}{\text{MPa}}\right)(5.34 \times 10^{-7} \text{ m}^4)}{0.025 \text{ m}}$$
$$= 2349.9 \text{ N·m} \quad (2300 \text{ N·m})$$

The answer is (C).

6. Convert degrees to radians.

$$\phi = (4.5°)\left(\frac{2\pi \text{ rad}}{360°}\right)$$
$$= 7.854 \times 10^{-2} \text{ rad}$$

The polar moment of inertia is

$$J = \frac{\pi r^4}{2} = \left(\frac{\pi}{2}\right)\left(\frac{25 \text{ mm}}{(2)\left(1000 \frac{\text{mm}}{\text{m}}\right)}\right)^4$$
$$= 3.83 \times 10^{-8} \text{ m}^4$$

Rearrange the twist angle equation to solve for torque.

$$T = \frac{\phi G J}{L}$$
$$= \frac{(7.854 \times 10^{-2} \text{ rad})(2.8 \times 10^{10} \text{ Pa})(3.83 \times 10^{-8} \text{ m}^4)}{\dfrac{50 \text{ cm}}{100 \frac{\text{cm}}{\text{m}}}}$$
$$= 168.7 \text{ N·m} \quad (170 \text{ N·m})$$

The answer is (D).

7. The elongation due to temperature change is

$$\delta = \alpha L(T_2 - T_1)$$
$$= \left(9.4 \times 10^{-6} \frac{1}{°\text{C}}\right)(1000 \text{ mm})(40°\text{C} - 10°\text{C})$$
$$= 0.282 \text{ mm}$$

Rearrange the elongation equation to solve for force.

$$F = \frac{\delta E A}{L}$$
$$= \frac{(0.282 \text{ mm})(200 \text{ GPa})\left(10^6 \frac{\text{kPa}}{\text{GPa}}\right)(2600 \text{ mm}^2)}{(1 \text{ m})\left(1000 \frac{\text{mm}}{\text{m}}\right)^3}$$
$$= 146.6 \text{ kN} \quad (150 \text{ kN})$$

The answer is (D).

8. Maximum shear stress occurs at the outer surface. The shear is

$$\tau = \frac{Tr}{J} = \frac{T\left(\dfrac{d}{2}\right)}{\dfrac{\pi}{32}d^4} = \frac{(280 \text{ N·m})\left(\dfrac{3 \text{ m}}{2}\right)}{\left(\dfrac{\pi}{32}\right)(3 \text{ m})^4}$$
$$= 52.8 \text{ Pa} \quad (53 \text{ Pa})$$

The answer is (D).

9. The thermal strain is

$$\varepsilon_t = \alpha \Delta T = \left(11.7 \times 10^{-6} \frac{1}{°\text{C}}\right)(50°\text{C})$$
$$= 0.000585 \text{ m/m}$$

The thermal stress is

$$\sigma_t = E \varepsilon_t = (210 \text{ GPa})\left(10^9 \frac{\text{Pa}}{\text{GPa}}\right)\left(0.000585 \frac{\text{m}}{\text{m}}\right)$$
$$= 1.2285 \times 10^8 \text{ Pa}$$

(This is less than the yield strength of steel.)

The compressive force in the rod is

$$F = \sigma A$$
$$= (1.2285 \times 10^8 \text{ Pa})\pi \left(\frac{12.5 \text{ mm}}{(2)\left(1000 \frac{\text{mm}}{\text{m}}\right)}\right)^2$$
$$= 15\,076 \text{ N} \quad (15 \text{ kN})$$

The answer is (B).

10. The total change in length is

$$\delta_t = \alpha L_{\text{initial}}(T - T_o)$$
$$= \left(11 \times 10^{-6} \frac{1}{°\text{C}}\right)(10 \text{ km})(50°\text{C} - 20°\text{C})$$
$$= 0.0033 \text{ km}$$

Add the change in length to the initial length.

$$L = L_{\text{initial}} + \delta_t$$
$$= 10 \text{ km} + 0.0033 \text{ km}$$
$$= 10.0033 \text{ km}$$

The answer is (C).

11. Tanks under external pressure fail by buckling (i.e., collapse), not by yielding. They should not be designed using the simplistic formulas commonly used for thin-walled tanks under internal pressure.

The answer is (D).

12. The torsional shear stress is maximum at the outer surface and is the same everywhere on the tube. The maximum moment occurs at the built-in end, tensile at the upper surface and compressive at the lower surface. The absolute value of the combined stress at the upper and lower surfaces at the built-in end will be the same.

The answer is (D).

13. Calculate the torque.

$$T = rF = \left(\frac{40 \text{ cm}}{(2)\left(100 \frac{\text{cm}}{\text{m}}\right)}\right)(45 \text{ kN})\left(1000 \frac{\text{N}}{\text{kN}}\right)$$
$$= 9000 \text{ N·m}$$

The polar moment of inertia is

$$J = \frac{\pi}{2}(r_o^4 - r_i^4)$$
$$= \left(\frac{\pi}{2}\right)\left(\left(\frac{10 \text{ cm}}{(2)\left(100 \frac{\text{cm}}{\text{m}}\right)}\right)^4 - \left(\frac{7.5 \text{ cm}}{(2)\left(100 \frac{\text{cm}}{\text{m}}\right)}\right)^4\right)$$
$$= 6.71 \times 10^{-6} \text{ m}^4$$

Find the angle of twist.

$$\phi = \frac{TL}{GJ} = \frac{(9000 \text{ N·m})(120 \text{ cm})}{(2.8 \times 10^{10} \text{ Pa})(6.71 \times 10^{-6} \text{ m}^4)\left(100 \frac{\text{cm}}{\text{m}}\right)}$$
$$= 0.057 \text{ rad}$$

Find the shear stress in the shaft.

$$\tau = \frac{Tr}{J} = \frac{(9000 \text{ N·m})\left(\frac{10 \text{ cm}}{(2)\left(100 \frac{\text{cm}}{\text{m}}\right)}\right)}{6.71 \times 10^{-6} \text{ m}^4}$$
$$= 67.05 \times 10^6 \text{ Pa} \quad (67 \text{ MPa})$$

The answer is (C).

14. Assume a thin-walled tank. Solve the equation for tangential (hoop) stress for the wall thickness. Although the inner radius is used by convention, the outer radius can be used.

$$t = \frac{pd}{2\sigma_t} = \frac{p(d_o - 2t)}{2\sigma_t} \approx \frac{pd_o}{2\sigma_t}$$
$$= \frac{(8 \text{ MPa})(25 \text{ cm})}{(2)(90 \text{ MPa})}$$
$$= 1.11 \text{ cm} \quad (1.1 \text{ cm})$$

Check the thin-wall assumption.

$$\frac{t}{r_i} = \frac{t}{\frac{d_o - 2t}{2}} = \frac{1.11 \text{ cm}}{\frac{25 \text{ cm} - (2)(1.11 \text{ cm})}{2}}$$
$$= 0.098 < 0.1 \quad [\text{thin wall}]$$

The answer is (C).

31 Beams

PRACTICE PROBLEMS

1. For the beam shown, what is most nearly the maximum compressive stress at section D-D, 1.5 m from the left end?

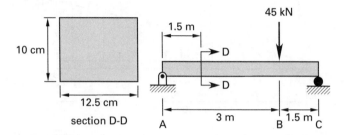

(A) 63 MPa

(B) 110 MPa

(C) 230 MPa

(D) 330 MPa

2. Refer to the beam shown. The beam is fixed at one end. The beam has a mass of 46.7 kg/m. The modulus of elasticity of the beam is 200 GPa; the moment of inertia is 4680 cm^4.

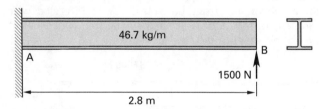

The upward force at B is 1500 N. What is most nearly the net deflection of the beam at a point 1.2 m from the fixed end?

(A) −0.32 mm (downward)

(B) −0.29 mm (downward)

(C) 0.12 mm (upward)

(D) 0.17 mm (upward)

3. Refer to the simply supported beam shown.

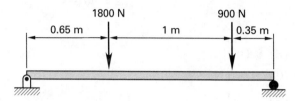

What is most nearly the maximum bending moment?

(A) 340 N·m

(B) 460 N·m

(C) 660 N·m

(D) 890 N·m

4. Refer to the cantilevered structural section shown. The beam is manufactured from steel with a modulus of elasticity of 210 GPa. The beam's cross-sectional area is 37.9 cm^2; its moment of inertia is 2880 cm^4. The beam has a mass of 45.9 kg/m. A 6000 N compressive force is applied at the top of the beam, at an angle of 30° from the horizontal. Neglect buckling.

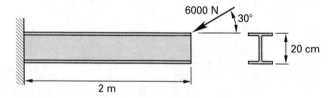

What is most nearly the maximum shear force in the beam?

(A) 3000 N

(B) 3900 N

(C) 5200 N

(D) 6100 N

5. For the cantilever steel rod shown, what is most nearly the force, F, necessary to deflect the rod a vertical distance of 7.5 mm?

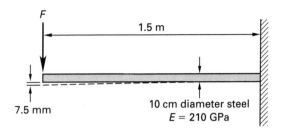

(A) 6900 N

(B) 8800 N

(C) 11 000 N

(D) 17 000 N

6. Refer to the simply supported beam shown.

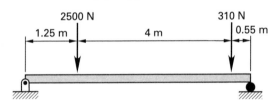

What is most nearly the maximum shear?

(A) 500 N

(B) 1000 N

(C) 1500 N

(D) 2000 N

7. Refer to the cantilevered structural section shown. The beam is manufactured from steel with a modulus of elasticity of 200 GPa. The beam's cross-sectional area is 74 cm², its moment of inertia is 8700 cm⁴. The beam has a mass of 60 kg/m. A 2500 N compressive force is applied at the top of the beam, at an angle of 22° from horizontal. Neglect buckling.

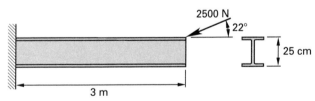

What is most nearly the approximate absolute value of the maximum bending moment in the beam?

(A) 5000 N·m

(B) 5200 N·m

(C) 5900 N·m

(D) 6100 N·m

8. A rectangular beam has a cross section of 5 cm wide × 10 cm deep and experiences a maximum shear of 2250 N. What is most nearly the maximum shear stress in the beam?

(A) 450 kPa

(B) 570 kPa

(C) 680 kPa

(D) 790 kPa

9. A simply supported beam supports a triangular distributed load as shown. The peak load at the right end of the beam is 5 N/m.

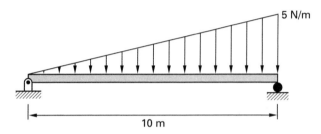

What is the approximate bending moment at a point 7 m from the left end of the beam?

(A) 15 N·m

(B) 17 N·m

(C) 28 N·m

(D) 30 N·m

10. Refer to the cantilevered structural section shown. The beam is manufactured from steel with a modulus of elasticity of 205 GPa. The beam's cross-sectional area is 86 cm²; its moment of inertia is 24 400 cm⁴. A 37 000 N compressive force is applied at the top of the beam, at an angle of 40° from horizontal. A counterclockwise moment of 600 N·m is applied to the free end. Neglect beam self-weight, and neglect buckling.

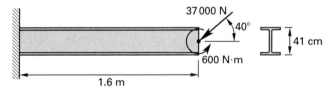

What is most nearly the deflection at the tip of the beam due to the external force alone (i.e., neglecting the beam's own mass)?

(A) 0.63 mm

(B) 0.82 mm

(C) 1.2 mm

(D) 2.5 mm

SOLUTIONS

1. Find the reaction at A.

$$\sum M_C = R_A(4.5 \text{ m}) - (45 \text{ kN})(1.5 \text{ m}) = 0$$
$$R_A = 15 \text{ kN}$$

The bending moment at section D-D is

$$M = (15 \text{ kN})(1.5 \text{ m}) = 22.5 \text{ kN·m}$$

The maximum compressive stress is at the top fiber of the beam section.

$$\sigma_{\max} = \frac{Mc}{I} = \frac{M\frac{h}{2}}{\frac{bh^3}{12}}$$

$$= \frac{(22.5 \text{ kN·m})\left(1000 \frac{\text{N}}{\text{kN}}\right)\left(\frac{0.10 \text{ m}}{2}\right)}{\frac{(0.125 \text{ m})(0.10 \text{ m})^3}{12}}$$

$$= 108 \times 10^6 \text{ Pa} \quad (110 \text{ MPa})$$

The answer is (B).

2. Use the principle of superposition to determine the deflection. The total deflection is the upward deflection due to the concentrated force less the downward deflection due to the weight of the beam.

Distance x is measured from the fixed end. The upward deflection due to the concentrated force is

$$v_{x,1} = \left(\frac{Px^2}{6EI}\right)(3L - x)$$

$$= \frac{(1500 \text{ N})(1.2 \text{ m})^2 \left(100 \frac{\text{cm}}{\text{m}}\right)^4}{(6)(200 \text{ GPa})\left(10^9 \frac{\text{Pa}}{\text{GPa}}\right)(4680 \text{ cm}^4)}$$

$$\times ((3)(2.8 \text{ m}) - 1.2 \text{ m})$$

$$= 0.000277 \text{ m} \quad (0.28 \text{ mm}) \quad [\text{upward}]$$

The downward deflection is due to the beam's own mass. Distance x is measured from the fixed end. The load per unit length is

$$w = mg = \left(46.7 \frac{\text{kg}}{\text{m}}\right)\left(9.81 \frac{\text{m}}{\text{s}^2}\right) = 458 \text{ N/m}$$

The downward deflection is

$$v_{x,2} = \left(\frac{-wx^2}{24EI}\right)(x^2 - 4Lx + 6L^2)$$

$$= \frac{-\left(458 \frac{\text{N}}{\text{m}}\right)(1.2 \text{ m})^2 \left(100 \frac{\text{cm}}{\text{m}}\right)^4}{(24)(200 \text{ GPa})\left(10^9 \frac{\text{Pa}}{\text{GPa}}\right)(4680 \text{ cm}^4)}$$

$$\times \left(\begin{array}{c}(1.2 \text{ m})^2 - (4)(2.8 \text{ m})(1.2 \text{ m}) \\ + (6)(2.8 \text{ m})^2\end{array}\right)$$

$$= -0.000103 \text{ m} \quad (-0.10 \text{ mm}) \quad [\text{downward}]$$

The net deflection is

$$v = v_{x,1} + v_{x,2} = 0.28 \text{ mm} + (-0.10 \text{ mm})$$
$$= 0.17 \text{ mm} \quad [\text{upward}]$$

The answer is (D).

3. Determine the reactions by taking moments about each end.

$$\sum M_B = -R_A(0.65 \text{ m} + 1 \text{ m} + 0.35 \text{ m})$$
$$+ (1800 \text{ N})(1 \text{ m} + 0.35 \text{ m})$$
$$+ (900 \text{ N})(0.35 \text{ m}) = 0$$
$$R_A = 1372.5 \text{ N}$$
$$\sum F_y = R_B + 1372.5 \text{ N} - 1800 \text{ N} - 900 \text{ N} = 0$$
$$R_B = 1327.5 \text{ N}$$

Draw the shear and moment diagrams.

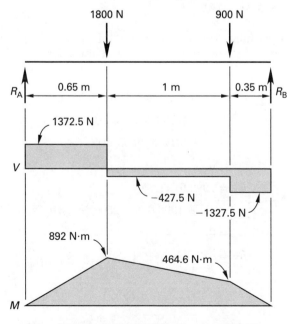

$$M = VR_A = (1372.5 \text{ N})(0.65 \text{ m}) = 892 \text{ N·m}$$

The maximum moment occurs 0.65 m from the left end (where V goes through zero) of the beam and is equal to 892 N·m (890 N·m).

The answer is (D).

4. The maximum vertical shear in the beam will occur at the fixed end.

$$V = wL + F_y$$
$$= mgL + F_y$$
$$= \left(45.9 \; \frac{\text{kg}}{\text{m}}\right)\left(9.81 \; \frac{\text{m}}{\text{s}^2}\right)(2 \text{ m}) + (6000 \text{ N})(\sin 30°)$$
$$= 3900 \text{ N}$$

The answer is (B).

5. For a cantilever beam loaded at its tip, with $x = L$,

$$v_{\max} = \frac{-PL^3}{3EI}$$
$$P = \frac{-3EIv_{\max}}{L^3}$$
$$= \frac{-(3)(210 \text{ GPa})\left(10^9 \; \frac{\text{Pa}}{\text{GPa}}\right) \times \left(\frac{\pi}{4}\right)(0.05 \text{ m})^4 (7.5 \text{ mm})}{(1.5 \text{ m})^3 \left(1000 \; \frac{\text{mm}}{\text{m}}\right)}$$
$$= -6872 \text{ N} \quad (6900 \text{ N}) \quad [\text{downward}]$$

The answer is (A).

6. Determine the reactions by taking the moments about end B and by taking the sum of the forces.

$$\sum M_B = -R_A(1.25 \text{ m} + 4 \text{ m} + 0.55 \text{ m})$$
$$+ (2500 \text{ N})(4 \text{ m} + 0.55 \text{ m})$$
$$+ (310 \text{ N})(0.55 \text{ m})$$
$$= 0$$
$$R_A = 1990.6 \text{ N}$$
$$\sum F_y = R_B + 1990.6 \text{ N} - 2500 \text{ N} - 310 \text{ N}$$
$$= 0$$
$$R_B = 819.4 \text{ N}$$

Draw the shear diagram.

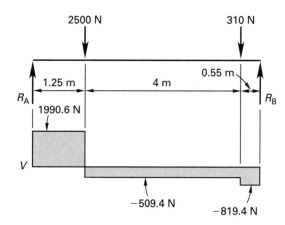

From the shear diagram, the maximum shear is 1990.6 N (2000 N).

The answer is (D).

7. The maximum bending moment will occur at the fixed end of the beam. The moment will be affected by the distributed load and the external force. Since the force does not act through the centroid of the beam (i.e., the force is eccentric), both the vertical and the horizontal components of the external force must be included.

The moment due to the beam's own mass is

$$M_1 = -\tfrac{1}{2}wL^2 = -\tfrac{1}{2}mgL^2$$
$$= -\left(\tfrac{1}{2}\right)\left(60 \; \frac{\text{kg}}{\text{m}}\right)\left(9.81 \; \frac{\text{m}}{\text{s}^2}\right)(3 \text{ m})^2$$
$$= -2648.7 \text{ N·m}$$

The moment due to the vertical component of the external force is

$$M_2 = -F_y L = -(2500 \text{ N})(\sin 22°)(3 \text{ m})$$
$$= -2809.5 \text{ N·m}$$

The force is not applied through the beam's centroid. The horizontal component of the force causes the beam to bend upward, while the other forces bend the beam downward. The moment due to the eccentricity is

$$M_3 = F_x e = (2500 \text{ N})(\cos 22°)\left(\frac{25 \text{ cm}}{(2)\left(100 \; \frac{\text{cm}}{\text{m}}\right)}\right)$$
$$= 289.7 \text{ N·m}$$

The total moment is

$$M = M_1 + M_2 + M_3$$
$$= -2648.7 \text{ N·m} - 2809.5 \text{ N·m} + 298.7 \text{ N·m}$$
$$= -5168.5 \text{ N·m} \quad (5200 \text{ N·m})$$

The answer is (B).

8. The maximum shear stress is

$$\tau_{max} = \frac{3V}{2A} = \frac{(3)(2250 \text{ N})\left(100 \frac{\text{cm}}{\text{m}}\right)^2}{(2)(5 \text{ cm})(10 \text{ cm})}$$
$$= 675 \times 10^3 \text{ Pa} \quad (680 \text{ kPa})$$

The answer is (C).

9. The total force from the distributed load is

$$\left(\frac{1}{2}\right)(10 \text{ m})\left(5 \frac{\text{N}}{\text{m}}\right) = 25 \text{ N}$$

This force can be assumed to act at two-thirds of the beam length from the left end, or one-third of the beam length from the right end.

Sum the moments around the right end to find the left reaction.

$$\sum M_{\text{right end}} = (25 \text{ N})\left(\frac{10 \text{ m}}{3}\right) - R_{\text{left}}(10 \text{ m}) = 0$$
$$R_{\text{left}} = 8.33 \text{ N}$$

The load increases linearly to 5 N/m at 10 m. At 7 m, the loading is $(0.7)(5 \text{ N/m})$. The total distributed force over the first 7 m of the beam is

$$\left(\frac{1}{2}\right)(7 \text{ m})\left((0.7)\left(5 \frac{\text{N}}{\text{m}}\right)\right) = 12.25 \text{ N}$$

Sum moments from the point of interest (7 m from the left end) to either end. The calculation is easier from the left end.

$$\sum M = (12.25 \text{ N})\left(\frac{7 \text{ m}}{3}\right) - (8.33 \text{ N})(7 \text{ m})$$
$$= -29.73 \text{ N·m} \quad (30 \text{ N·m})$$

The answer is (D).

10. With $x = L$, the deflection due to the vertical component of the force is

$$v_1 = \frac{-PL^3}{3EI} = \frac{-(37\,000 \text{ N})(\sin 40°)(1.6 \text{ m})^3 \left(100 \frac{\text{cm}}{\text{m}}\right)^4}{(3)(205 \text{ GPa})\left(10^9 \frac{\text{Pa}}{\text{GPa}}\right)(24\,400 \text{ cm}^4)}$$
$$= -0.000649 \text{ m} \quad (-0.649 \text{ mm}) \quad [\text{downward}]$$

The eccentric application of the force causes an upward deflection. The deflection due to the end moment is

$$v_2 = \frac{M_0 L^2}{2EI} = \frac{(600 \text{ N·m})(1.6 \text{ m})^2 \left(100 \frac{\text{cm}}{\text{m}}\right)^4}{(2)(205 \text{ GPa})\left(10^9 \frac{\text{Pa}}{\text{GPa}}\right)(24\,400 \text{ cm}^4)}$$
$$= 0.0000154 \text{ m} \quad (0.0154 \text{ mm}) \quad [\text{upward}]$$

The total deflection due to the external force alone is

$$v = v_1 + v_2 = -0.649 \text{ mm} + 0.0154 \text{ mm}$$
$$= -0.634 \text{ mm} \quad (0.63 \text{ mm}) \quad [\text{downward}]$$

The answer is (A).

32 Columns

PRACTICE PROBLEMS

1. A steel column with a cross section of 12 cm × 16 cm is 4 m in height and fixed at its base. The column is pinned against translation in its weak direction at the top but is unbraced in its strong direction. The column's modulus of elasticity is 2.1×10^5 MPa.

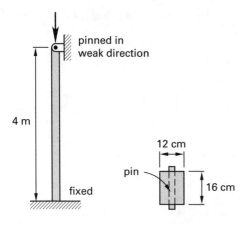

What is most nearly the maximum theoretical vertical load the column can support without buckling?

(A) 1.3 MN

(B) 5.2 MN

(C) 6.1 MN

(D) 11 MN

2. A 10 cm × 10 cm square column supports a compressive force of 9000 N. The load is applied with an eccentricity of 2.5 cm along one of the lines of symmetry. What is most nearly the maximum tensile stress in the column?

(A) 450 kPa

(B) 900 kPa

(C) 1400 kPa

(D) 2300 kPa

3. A square column with a solid cross section is placed in a building to support a load of 5 MN. The maximum allowable stress in the column is 350 MPa. The column reacts linearly to all loads. If the contractor is permitted to load the column anywhere in the central one-fifth of the column's cross section, what are most nearly the smallest possible dimensions of the column?

(A) 12 cm × 12 cm

(B) 14 cm × 14 cm

(C) 16 cm × 16 cm

(D) 18 cm × 18 cm

4. What is most nearly the maximum resultant normal stress at A for the cantilever beam shown?

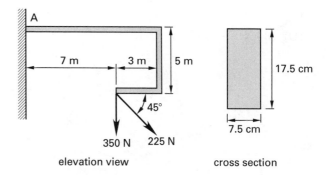

(A) 7.2 MPa

(B) 9.4 MPa

(C) 9.8 MPa

(D) 9.9 MPa

5. A rectangular steel bar 37.5 mm wide and 50 mm thick is pinned at each end and subjected to axial compression. The bar has a length of 1.75 m. The modulus of elasticity is 200 GPa. What is most nearly the critical buckling load?

(A) 60 kN

(B) 93 kN

(C) 110 kN

(D) 140 kN

6. What is most nearly the Euler buckling load for a 10 m long steel column with unrestrained ends and with the given properties and cross section?

$$I_{x'x'} = 3.70 \times 10^6 \text{ mm}^4$$

$$E = 200 \text{ GPa}$$

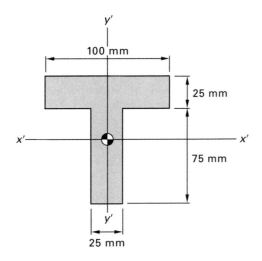

(A) 15 kN
(B) 24 kN
(C) 43 kN
(D) 73 kN

SOLUTIONS

1. Since the column is fixed at one end and pinned at the other, the theoretical end-restraint coefficient, K, is 0.7. The effective length for buckling in the weak direction is

$$K\ell = (0.7)(4 \text{ m}) = 2.8 \text{ m}$$

The moment of inertia for buckling in the weak direction is

$$I = \frac{bh^3}{12} = \frac{(16 \text{ cm})(12 \text{ cm})^3}{(12)\left(100 \dfrac{\text{cm}}{\text{m}}\right)^4}$$

$$= 2.3 \times 10^{-5} \text{ m}^4$$

Calculate the critical buckling force from Euler's formula.

$$P_{cr} = \frac{\pi^2 EI}{(K\ell)^2}$$

$$= \frac{\pi^2 (2.1 \times 10^5 \text{ MPa})\left(10^6 \dfrac{\text{Pa}}{\text{MPa}}\right)(2.3 \times 10^{-5} \text{ m}^4)}{(2.8 \text{ m})^2}$$

$$= 6.09 \times 10^6 \text{ N} \quad (6.1 \text{ MN})$$

Check the buckling force in the strong direction. The column is not braced in that direction, so for a column fixed at one end and free at the other, $K = 2$.

$$K\ell = (2)(4 \text{ m}) = 8 \text{ m}$$

The moment of inertia for buckling in the strong direction is

$$I = \frac{bh^3}{12} = \frac{(12 \text{ cm})(16 \text{ cm})^3}{(12)\left(100 \dfrac{\text{cm}}{\text{m}}\right)^4} = 4.1 \times 10^{-5} \text{ m}^4$$

Calculate the critical buckling force from Euler's formula.

$$P_{cr} = \frac{\pi^2 EI}{(K\ell)^2}$$

$$= \frac{\pi^2 (2.1 \times 10^5 \text{ MPa})\left(10^6 \dfrac{\text{Pa}}{\text{MPa}}\right)(4.1 \times 10^{-5} \text{ m}^4)}{(8 \text{ m})^2}$$

$$= 1.3 \times 10^6 \text{ N} \quad (1.3 \text{ MN})$$

This is less than for buckling in the weak direction. This force controls.

The answer is (A).

2. The cross-sectional area of a square column is

$$A = b^2 = \left(\frac{10 \text{ cm}}{100 \frac{\text{cm}}{\text{m}}}\right)^2 = 0.01 \text{ m}^2$$

The moment of inertia of the square cross section is

$$I = \frac{b^4}{12} = \frac{(10 \text{ cm})^4}{(12)\left(100 \frac{\text{cm}}{\text{m}}\right)^4} = 8.33 \times 10^{-6} \text{ m}^4$$

The distance from the neutral axis to the extreme fibers is

$$c = \frac{b}{2} = \frac{10 \text{ cm}}{(2)\left(100 \frac{\text{cm}}{\text{m}}\right)} = 0.05 \text{ m}$$

The stress is

$$\sigma = \frac{F}{A} \pm \frac{Fec}{I}$$

$$= \frac{-9000 \text{ N}}{0.01 \text{ m}^2} \pm \frac{(-9000 \text{ N})(2.5 \text{ cm})(0.05 \text{ m})}{(8.33 \times 10^{-6} \text{ m}^4)\left(100 \frac{\text{cm}}{\text{m}}\right)}$$

$$= -9 \times 10^5 \text{ Pa} \pm 1.35 \times 10^6 \text{ Pa}$$

$$(-900 \text{ kPa} \pm 1350 \text{ kPa})$$

The first term is due to the compressive column load and is compressive. (Compressive forces and stresses are usually given a negative sign.) The second term is due to the eccentricity. The second term increases the compressive stress at the inner face. It counteracts the compressive stress at the outer face.

The maximum tensile stress is

$$\sigma_{t,\text{max}} = -900 \text{ kPa} + 1350 \text{ kPa} = 450 \text{ kPa}$$

The answer is (A).

3. The middle one-fifth of the column is a square with dimensions of $b/5 \times b/5$ $(0.2b \times 0.2b)$.

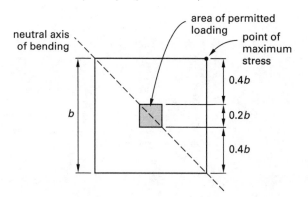

The maximum stress will be induced when the middle one-fifth square is loaded at one of its corners.

The cross-sectional area is

$$A = b^2$$

The moment of inertia of the square cross section is

$$I = \frac{b^4}{12}$$

The distance from the neutral axis to the extreme fibers is

$$c = \frac{b}{2}$$

The maximum eccentricity is

$$e = 0.1b$$

The stress at the extreme corner is

$$\sigma = \frac{F}{A} \pm \frac{Fe_x c_x}{I_x} + \frac{Fe_y c_y}{I_y}$$

$$= F\left(\frac{1}{b^2} \pm \frac{(2)(0.1b)\left(\frac{b}{2}\right)}{\frac{b^4}{12}}\right)$$

$$= F\left(\frac{1}{b^2} \pm \frac{1.2}{b^2}\right)$$

$$= \frac{2.2F}{b^2}$$

$$b = \sqrt{\frac{2.2F}{\sigma}}$$

$$= \sqrt{\frac{(2.2)(5 \text{ MN})\left(10^6 \frac{\text{N}}{\text{MN}}\right)}{(350 \text{ MPa})\left(10^6 \frac{\text{Pa}}{\text{MPa}}\right)}}$$

$$= 0.177 \text{ m} \quad (18 \text{ cm})$$

The answer is (D).

4. The beam experiences both axial tension and bending stresses, so it should be analyzed as a beam-column.

$$\sum M_A = (350 \text{ N})(7 \text{ m}) + (225 \text{ N})(\sin 45°)(7 \text{ m})$$
$$- (225 \text{ N})(\cos 45°)(5 \text{ m})$$
$$= 2768 \text{ N·m}$$

The stress is

$$\sigma_{max} = \frac{P}{A} + \frac{Mc}{I} = \frac{P}{bh} + \frac{M\left(\frac{b}{2}\right)}{\frac{bh^3}{12}}$$

$$= \frac{(225 \text{ N})(\cos 45°)\left(100 \frac{\text{cm}}{\text{m}}\right)^2}{(7.5 \text{ cm})(17.5 \text{ cm})}$$

$$+ \frac{(2768 \text{ N·m})\left(\frac{17.5 \text{ cm}}{(2)\left(100 \frac{\text{cm}}{\text{m}}\right)}\right)}{\frac{(7.5 \text{ cm})(17.5 \text{ cm})^3}{(12)\left(100 \frac{\text{cm}}{\text{m}}\right)^4}}$$

$$= 7.24 \times 10^6 \text{ Pa} \quad (7.2 \text{ MPa})$$

The answer is (A).

5. Use Euler's formula. $K = 1$ since both ends are pinned. Use the moment of inertia for the weak direction.

$$P_{cr} = \frac{\pi^2 EI}{(K\ell)^2} = \frac{\pi^2 E\left(\frac{bh^3}{12}\right)}{(K\ell)^2}$$

$$= \frac{\pi^2 (200 \text{ GPa})\left(10^9 \frac{\text{Pa}}{\text{GPa}}\right)\left(\frac{(50 \text{ mm})(37.5 \text{ mm})^3}{(12)\left(1000 \frac{\text{mm}}{\text{m}}\right)^4}\right)}{((1)(1.75 \text{ m}))^2}$$

$$= 141\,624 \text{ N} \quad (140 \text{ kN})$$

The answer is (D).

6. $x'x'$ and $y'y'$ are centroidal axes. $I_{y'y'}$ is computed from the equation $I = bh^3/12$ about the centroidal axis of a rectangle. For this cross section, $b_1 = 25$ mm, $h_1 = 100$ mm, $b_2 = 75$ mm, and $h_2 = 25$ mm.

$$I_{y'y'} = \frac{b_1 h_1^3}{12} + \frac{b_2 h_2^3}{12}$$

$$= \frac{(25 \text{ mm})(100 \text{ mm})^3}{(12)\left(1000 \frac{\text{mm}}{\text{m}}\right)^4} + \frac{(75 \text{ mm})(25 \text{ mm})^3}{(12)\left(1000 \frac{\text{mm}}{\text{m}}\right)^4}$$

$$= 2.18 \times 10^{-6} \text{ m}^4$$

Find the Euler buckling load, P_{cr}. The smallest moment of inertia (corresponding to the least radius of gyration) should be used. $I_{y'y'}$ is less than $I_{x'x'}$.

$$P_{cr} = \frac{\pi^2 EI}{(K\ell)^2}$$

$$= \frac{\pi^2 (200 \text{ GPa})\left(10^6 \frac{\text{kPa}}{\text{GPa}}\right)(2.18 \times 10^{-6} \text{ m}^4)}{((1)(10 \text{ m}))^2}$$

$$= 43 \text{ kN}$$

The answer is (C).

33 Electrostatics

PRACTICE PROBLEMS

1. A 0.001 C charge is separated from a 0.003 C charge by 10 m. Point P identifies the point of zero electric field between the charges.

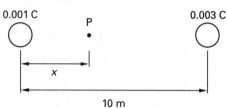

What is most nearly the distance, x, between the 0.001 C charge and point P?

(A) 2.2 m
(B) 3.7 m
(C) 6.3 m
(D) 14 m

2. A 15 μC point charge is located on the y-axis at $(0, 0.25)$. A second charge of 10 μC is located on the x-axis at $(0.25, 0)$.

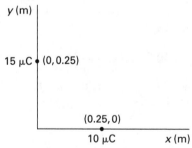

If the two charges are separated by air, what is most nearly the force between them?

(A) 0.098 N
(B) 0.34 N
(C) 11 N
(D) 34 N

3. A charge of 0.75 C passes through a wire every 15 s. What is most nearly the current in the wire?

(A) 5.0 mA
(B) 10 mA
(C) 20 mA
(D) 50 mA

4. If a potential difference is generated by a single conductor passing through a magnetic field, which statement is FALSE?

(A) The potential difference depends on the speed with which the conductor cuts the magnetic field.
(B) The potential difference depends on the length of the conductor that cuts the magnetic field.
(C) The potential difference depends on the magnetic field density that is present.
(D) The potential difference depends on the diameter of the conductor that cuts the magnetic field.

5. A current of 10 A flows through a 1 mm diameter wire. What is most nearly the average number of electrons per second that pass through a cross section of the wire?

(A) 1.6×10^{18} electrons/s
(B) 6.2×10^{18} electrons/s
(C) 1.6×10^{19} electrons/s
(D) 6.3×10^{19} electrons/s

6. Point charges Q_1, Q_2, and Q_3 are arranged as shown.

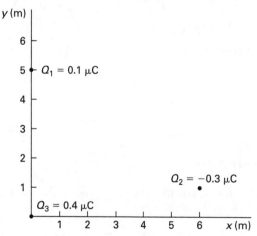

What is most nearly the magnitude of the force on Q_3 due to Q_1 and Q_2?

(A) 2.3×10^{-5} N
(B) 3.0×10^{-5} N
(C) 9.8×10^{-4} N
(D) 5.1×10^{-2} N

SOLUTIONS

1. At the point where the electric field intensity E at point 2 is zero, the electric field due to the 0.001 C charge equals the field due to the 0.003 C charge in magnitude. r is the distance between point 1 and point 2.

$$E = \frac{Q_1}{4\pi\varepsilon r^2}$$

$$\frac{0.001 \text{ C}}{4\pi\varepsilon_0 x^2} = \frac{0.003 \text{ C}}{4\pi\varepsilon_0 (10 \text{ m} - x)^2}$$

$$(10 \text{ m} - x)^2 = 3x^2$$

$$x^2 + 10x - 50 = 0$$

Use the quadratic equation to solve for the positive x value.

$$x = \frac{-b + \sqrt{b^2 - 4ac}}{2a}$$

$$= \frac{-10 \text{ m} + \sqrt{(10 \text{ m})^2 - (4)(1)(-50 \text{ m}^2)}}{(2)(1)}$$

$$= 3.66 \text{ m} \quad (3.7 \text{ m})$$

The answer is (B).

2. The radius is

$$r = \sqrt{(0.25 \text{ m})^2 + (0.25 \text{ m})^2} = \frac{\sqrt{2}}{4} \text{ m}$$

For air, $\varepsilon = 8.85 \times 10^{-12}$ F/m. The units F/m are equivalent to $C^2/N \cdot m^2$. Using Coulomb's law, the force between the two charges is

$$F = \frac{Q_1 Q_2}{4\pi\varepsilon r^2} = \frac{(15 \times 10^{-6} \text{ C})(10 \times 10^{-6} \text{ C})}{4\pi\left(8.85 \times 10^{-12} \frac{\text{F}}{\text{m}}\right)\left(\frac{\sqrt{2}}{4} \text{ m}\right)^2}$$

$$= 10.79 \text{ N} \quad (11 \text{ N})$$

The answer is (C).

3. Current is the charge per unit time passing through the wire.

$$I = \frac{q}{t} = \frac{(0.75 \text{ C})\left(1000 \frac{\text{mA}}{\text{A}}\right)}{15 \text{ s}} = 50 \text{ mA}$$

The answer is (D).

4. The potential difference is the induced voltage described by Faraday's law.

$$v = \frac{-N d\phi}{dt}$$

The change in magnetic flux, $d\phi/dt$, will be influenced by the length of the conductor but not by the cross-sectional area or diameter of the conductor. For a single conductor, $N=1$.

The answer is (D).

5. A current of 10 A is equivalent to 10 C/s. One electron has a charge of approximately 1.6×10^{-19} C.

$$\dot{q} = \frac{I}{Q} = \frac{10 \frac{\text{C}}{\text{s}}}{1.6 \times 10^{-19} \frac{\text{C}}{\text{electron}}}$$

$$= 6.25 \times 10^{19} \text{ electrons/s} \quad (6.3 \times 10^{19} \text{ electrons/s})$$

The wire diameter is irrelevant.

The answer is (D).

6. The force between Q_1 and Q_3 is repulsive and is entirely in the negative y-direction.

The distance between Q_1 and Q_3 is 5 m.

$$F_{3-1} = \frac{Q_1 Q_3}{4\pi\varepsilon r^2} = \frac{(0.1 \times 10^{-6} \text{ C})(0.4 \times 10^{-6} \text{ C})}{4\pi\left(8.85 \times 10^{-12} \frac{\text{F}}{\text{m}}\right)(5 \text{ m})^2}$$

$$= 1.44 \times 10^{-5} \text{ N}$$

The distance between Q_2 and Q_3 is

$$\sqrt{(1 \text{ m})^2 + (6 \text{ m})^2} = 6.08 \text{ m}$$

The force between Q_2 and Q_3 is attractive and is partly in the positive y-direction and partly in the positive x-direction.

$$F_{3-2} = \frac{Q_2 Q_3}{4\pi\varepsilon r^2} = \frac{(-0.3 \times 10^{-6} \text{ C})(0.4 \times 10^{-6} \text{ C})}{4\pi\left(8.85 \times 10^{-12} \frac{\text{F}}{\text{m}}\right)(6.08 \text{ m})^2}$$

$$= -2.92 \times 10^{-5} \text{ N}$$

The portion of the force between Q_2 and Q_3 in the positive y-direction is

$$(-2.92 \times 10^{-5} \text{ N})\left(\frac{1 \text{ m}}{6.08 \text{ m}}\right) = -4.8 \times 10^{-6} \text{ N}$$

The portion of the force between Q_2 and Q_3 in the positive x-direction is

$$(-2.92 \times 10^{-5} \text{ N})\left(\frac{6 \text{ m}}{6.08 \text{ m}}\right) = -2.88 \times 10^{-5} \text{ N}$$

The total force is the vector sum of the two forces.

$$F_{\text{total}} = \sqrt{\left(1.44 \times 10^{-5} \text{ N} + (-4.8 \times 10^{-6} \text{ N})\right)^2 + (-2.88 \times 10^{-5} \text{ N})^2}$$

$$= 3.04 \times 10^{-5} \text{ N} \quad (3.0 \times 10^{-5} \text{ N})$$

The answer is (B).

34 Direct-Current Circuits

PRACTICE PROBLEMS

1. What is most nearly the equivalent inductance of the circuit shown?

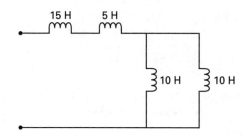

(A) 5.0 H
(B) 20 H
(C) 24 H
(D) 25 H

2. A solid copper conductor at 20°C has the following characteristics.

resistivity = 1.77×10^{-8} Ω·m
diameter = 5 mm
length = 5000 m

What is most nearly the resistance of the conductor?

(A) 0.017 Ω
(B) 4.5 Ω
(C) 12 Ω
(D) 18 Ω

3. The circuit shown is in steady state.

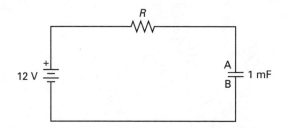

What is most nearly the charge on the capacitor on plate A?

(A) 83 pC
(B) 120 pC
(C) 83 μC
(D) 0.012 C

4. What is most nearly the current, I, in the illustration shown?

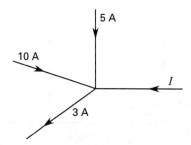

(A) −12 A
(B) 3.0 A
(C) 5.0 A
(D) 15 A

5. What is most nearly the voltage across the 5 Ω resistor in the center leg in the illustration shown?

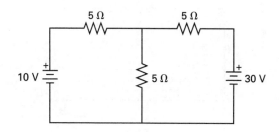

(A) 13 V
(B) 16 V
(C) 20 V
(D) 24 V

6. In the circuit shown, what is most nearly the current through CD?

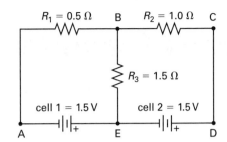

(A) 0.20 A
(B) 0.60 A
(C) 1.0 A
(D) 1.9 A

7. What are most nearly the Norton equivalent voltage, V_N, and resistance, R_N, values for the circuit shown?

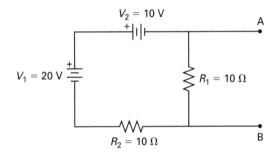

(A) $V_N = 5$ V; $R_N = 5\ \Omega$
(B) $V_N = 10$ V; $R_N = 20\ \Omega$
(C) $I_N = 1$ A; $R_N = 5\ \Omega$
(D) $I_N = 1$ A; $R_N = 10\ \Omega$

8. When a 20 V source is connected across terminals A and B, a current of 10 A is measured through R_1.

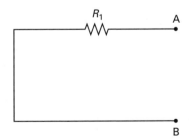

Approximately what current would flow through R_1 if a 30 V source is connected across terminals A and B?

(A) 12 A
(B) 15 A
(C) 17 A
(D) 20 A

9. What is most nearly the voltage drop across the 6 Ω resistor in the circuit shown?

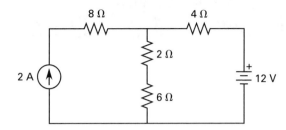

(A) 6.0 V
(B) 9.0 V
(C) 10 V
(D) 18 V

10. A current of 7 A passes through a 12 Ω resistor. What is most nearly the power dissipated in the resistor?

(A) 0.1 hp
(B) 0.6 hp
(C) 0.8 hp
(D) 8 hp

SOLUTIONS

1. Inductors combine like resistors.

$$L_{eq} = 15 \text{ H} + 5 \text{ H} + \frac{(10 \text{ H})(10 \text{ H})}{10 \text{ H} + 10 \text{ H}}$$
$$= 25 \text{ H}$$

The answer is (D).

2. Calculate the resistance.

$$R = \frac{\rho L}{A} = \frac{\rho L}{\frac{\pi}{4} d^2}$$
$$= \frac{(1.77 \times 10^{-8} \ \Omega \cdot \text{m})(5000 \text{ m})}{\left(\frac{\pi}{4}\right)\left(\frac{5 \text{ mm}}{1000 \frac{\text{mm}}{\text{m}}}\right)^2}$$
$$= 4.51 \ \Omega \quad (4.5 \ \Omega)$$

The answer is (B).

3. In steady state, all of the voltage is across the capacitor.

$$Q = CV = (1 \times 10^{-3} \text{ F})(12 \text{ V})$$
$$= 0.012 \text{ C}$$

The answer is (D).

4. Kirchhoff's current law states that the sum of the currents entering a junction will be equal to the sum of the currents leaving the junction. To apply the law, use assumed directions for currents.

$$\sum I_{in} = I + 5 \text{ A} + 10 \text{ A}$$
$$\sum I_{out} = 3 \text{ A}$$
$$\sum I_{in} = \sum I_{out}$$
$$I = 3 \text{ A} - 5 \text{ A} - 10 \text{ A}$$
$$= -12 \text{ A}$$

The answer is (A).

5. Use the loop-current method to solve for the voltage. This is a three-loop network, so select $(3-1)$, or 2, loops. Current directions are arbitrary.

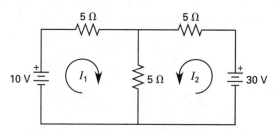

Write Kirchhoff's voltage law for each loop.

$$10 \text{ V} - I_1(5 \ \Omega) - (I_1 + I_2)(5 \ \Omega) = 0 \quad [\text{Eq. I}]$$
$$30 \text{ V} - I_2(5 \ \Omega) - (I_1 + I_2)(5 \ \Omega) = 0 \quad [\text{Eq. II}]$$

Solving for I_1 in Eq. I,

$$I_1 = \frac{10 \text{ V} - (5 \ \Omega)I_2}{10 \ \Omega}$$

Substituting into Eq. II,

$$30 \text{ V} - I_2(10 \ \Omega) - \left(\frac{10 \text{ V} - (5 \ \Omega)I_2}{10 \ \Omega}\right)(5 \ \Omega) = 0$$
$$I_2 = 10/3 \text{ A}$$
$$I_1 = -2/3 \text{ A}$$

The voltage across the center resistor is

$$V = IR = \left(\frac{10}{3} \text{ A} + \left(-\frac{2}{3} \text{ A}\right)\right)(5 \ \Omega)$$
$$= 13.33 \text{ V} \quad (13 \text{ V})$$

The answer is (A).

6. The method of superposition is used to find the current, I. I_1 is the current from cell 1, and I_2 is the current from cell 2. Separate the circuits.

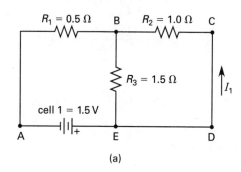
(a)

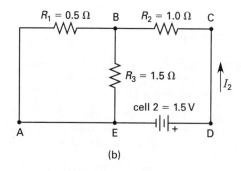
(b)

Short-circuiting cell 2 to find I_1 as shown in illustration (a), the equivalent total resistance is

$$R_{\text{total},1} = R_1 + \frac{R_2 R_3}{R_2 + R_3}$$
$$= 0.5 \ \Omega + \frac{(1.0 \ \Omega)(1.5 \ \Omega)}{1.0 \ \Omega + 1.5 \ \Omega}$$
$$= 1.1 \ \Omega$$

The current through cell 1 is

$$I_{\text{cell}\,1} = \frac{V}{R_{\text{total},1}} = \frac{1.5 \ \text{V}}{1.1 \ \Omega}$$

All of this cell 1 current passes through resistor R_1. The current then splits in inverse proportion to the leg resistances.

$$I_1 = I_{\text{cell}\,1} \left(\frac{R_3}{R_2 + R_3} \right)$$
$$= \left(\frac{1.5 \ \text{V}}{1.1 \ \Omega} \right) \left(\frac{1.5 \ \Omega}{1.0 \ \Omega + 1.5 \ \Omega} \right)$$
$$= 0.82 \ \text{A}$$

Short-circuiting cell 1 to find I_2, as shown in illustration (b), the equivalent total resistance is

$$R_{\text{total},2} = \frac{R_1 R_3}{R_1 + R_3} + R_2$$
$$= \frac{(0.5 \ \Omega)(1.5 \ \Omega)}{0.5 \ \Omega + 1.5 \ \Omega} + 1.0 \ \Omega$$
$$= 1.375 \ \Omega$$
$$I_2 = \frac{V}{R} = \frac{1.5 \ \text{V}}{1.375 \ \Omega}$$
$$= 1.1 \ \text{A}$$

The total current is

$$I = I_1 + I_2 = 0.82 \ \text{A} + 1.1 \ \text{A}$$
$$= 1.91 \ \text{A} \quad (1.9 \ \text{A})$$

The answer is (D).

7. A Norton equivalent circuit contains a single current source and a resistor in parallel, so the options A and B can be eliminated immediately.

To find the equivalent resistance, turn off all power sources (i.e., short-circuit the voltage sources). The equivalent resistance across terminals A and B is

$$R_N = \frac{R_1 R_2}{R_1 + R_2} = \frac{(10 \ \Omega)(10 \ \Omega)}{10 \ \Omega + 10 \ \Omega}$$
$$= 5 \ \Omega$$

(This eliminates options B and D.)

The Norton equivalent current is the short-circuit current through terminals A and B, as shown.

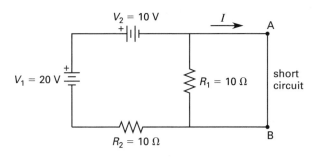

Apply Kirchhoff's voltage law to the shorted circuit.

$$V_1 - V_2 - R_2 I_N = 0$$
$$20 \ \text{V} - 10 \ \text{V} = (10 \ \Omega) I_N$$
$$I_N = 1 \ \text{A}$$

No current flows through R_1 because the terminals of R_1 are short-circuited.

The answer is (C).

8. The applied voltage increased from 20 V to 30 V, or 1.5 times. By the linearity expressed in Ohm's law, the current is

$$I = (1.5)(10 \ \text{A}) = 15 \ \text{A}$$

The answer is (B).

9. Redraw the circuit.

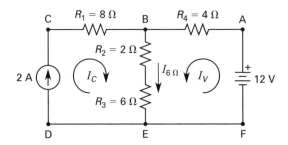

From superposition, with I_C designating the current through the resistor from the current source, and I_V designating the current through the resistor from the voltage source,

$$I_{6\Omega} = I_C + I_V$$

The current through the resistor is

$$I_C = (2 \ \text{A}) \left(\frac{R_4}{R_2 + R_3 + R_4} \right)$$
$$= (2 \ \text{A}) \left(\frac{4 \ \Omega}{2 \ \Omega + 6 \ \Omega + 4 \ \Omega} \right)$$
$$= 0.67 \ \text{A}$$

Use Kirchhoff's voltage law, and solve for I_V.

$$V - I_V(R_4 + R_2 + R_3) = 0$$

$$\begin{aligned}I_V &= \frac{V}{R_2 + R_3 + R_4} \\ &= \frac{12\text{ V}}{2\text{ }\Omega + 6\text{ }\Omega + 4\text{ }\Omega} \\ &= 1\text{ A}\end{aligned}$$

The voltage drop across the resistor is

$$I_{6\Omega} = I_C + I_V = 0.67\text{ A} + 1\text{ A} = 1.67\text{ A}$$
$$V_{6\Omega} = I_{6\Omega}R = (1.67\text{ A})(6\text{ }\Omega) = 10\text{ V}$$

The answer is (C).

10. The power dissipated through the resistor is

$$\begin{aligned}P = I^2 R &= \frac{(7\text{ A})^2(12\text{ }\Omega)}{745.7\text{ }\frac{\text{W}}{\text{hp}}} \\ &= 0.79\text{ hp} \quad (0.8\text{ hp})\end{aligned}$$

The answer is (C).

35 Alternating-Current Circuits

PRACTICE PROBLEMS

1. Which measurements are required to determine the phase angle of a single-phase circuit?

(A) power consumed by the circuit
(B) frequency, capacitance, and inductance
(C) power, voltage, and current
(D) resistance, current, and voltage

2. Which expression correctly relates power factor, pf, to real power, P, reactive power, Q, and complex power, S?

(A) $\text{pf} = \dfrac{P}{Q}$
(B) $\text{pf} = \dfrac{Q}{P}$
(C) $\text{pf} = \dfrac{P}{S}$
(D) $\text{pf} = \dfrac{Q}{S}$

3. To replace a heat loss of 2 kW, a 20°C room is heated by a resistive element heater from a standard 60 Hz power supply with an rms voltage of 120 V. Most nearly, what must the element's resistance be to maintain the room at 20°C?

(A) 4.6 Ω
(B) 7.2 Ω
(C) 14 Ω
(D) 17 Ω

4. What is most nearly the resonant frequency of the circuit shown?

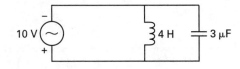

(A) 1.9 Hz
(B) 4.6 Hz
(C) 46 Hz
(D) 75 Hz

5. A 240 V alternating source at 60 Hz is connected to a series-RLC circuit as shown.

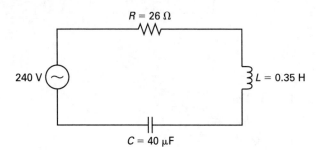

What is most nearly the total reactance of the circuit?

(A) 66 Ω
(B) 130 Ω
(C) 150 Ω
(D) 200 Ω

6. In the circuit shown, approximately what capacitance is needed to achieve a power factor of 1.0?

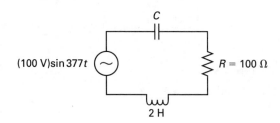

(A) 1.0 μF
(B) 1.6 μF
(C) 2.0 μF
(D) 3.5 μF

7. If the capacitor and the inductor in the circuit shown have the same reactance, what is most nearly the frequency of the AC source?

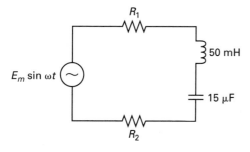

(A) 27 Hz
(B) 180 Hz
(C) 210 Hz
(D) 1200 Hz

8. What is most nearly the resonant frequency for the circuit shown?

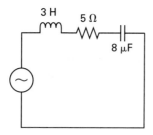

(A) 0 rad/s
(B) 0.2 rad/s
(C) 9 rad/s
(D) 200 rad/s

SOLUTIONS

1. The power is

$$P = \tfrac{1}{2} V_{max} I_{max} \cos\theta$$
$$= V_{rms} I_{rms} \cos\theta$$

To find the phase angle, θ, the power, voltage, and current must be known.

The answer is (C).

2. The power factor is defined as $\cos\theta$, where θ is the angle between the real and complex power vectors in the complex power triangle.

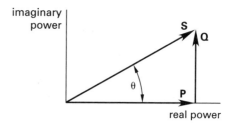

Standard trigonometric identities provide the relationship between θ and P, Q, and S.

$$\text{pf} = \cos\theta = \frac{\text{adjacent}}{\text{hypotenuse}} = \frac{P}{S}$$

The answer is (C).

3. The heater is a purely resistive element. The resistance can be solved directly by using the effective voltage.

$$P_{ave} = \frac{V_{rms}^2}{R}$$

$$R = \frac{V_{rms}^2}{P_{ave}} = \frac{(120\ \text{V})^2}{(2\ \text{kW})\left(1000\ \frac{\text{W}}{\text{kW}}\right)}$$

$$= 7.2\ \Omega$$

The answer is (B).

4. The resonant frequency is

$$\frac{1}{\sqrt{LC}} = 2\pi f_0$$

$$f_0 = \frac{1}{2\pi\sqrt{LC}} = \frac{1}{2\pi\sqrt{(4\ \text{H})(3\times 10^{-6}\ \text{F})}}$$

$$= 45.94\ \text{Hz}\quad (46\ \text{Hz})$$

The answer is (C).

5. The inductive reactance is

$$X_L = \omega L = 2\pi f L = 2\pi (60 \text{ Hz})(0.35 \text{ H})$$
$$= 131.9 \text{ }\Omega$$

The capacitive reactance is

$$X_C = \frac{1}{\omega C} = \frac{1}{2\pi f C} = \frac{1}{2\pi (60 \text{ Hz})(40 \times 10^{-6} \text{ F})}$$
$$= 66.3 \text{ }\Omega$$

The total reactance is

$$X_L - X_C = 131.9 \text{ }\Omega - 66.3 \text{ }\Omega$$
$$= 65.6 \text{ }\Omega \quad (66 \text{ }\Omega)$$

The answer is (A).

6. A circuit has a power factor of 1.0 when it is resonant. The input frequency is given as 377 rad/s. Size the capacitor so that the circuit resonates at the input frequency.

$$\omega_0 L = \frac{1}{\omega_0 C}$$
$$C = \frac{1}{\omega_0^2 L} = \frac{1}{\left(377 \frac{\text{rad}}{\text{s}}\right)^2 (2 \text{ H})}$$
$$= 3.52 \times 10^{-6} \text{ F} \quad (3.5 \text{ }\mu\text{F})$$

The answer is (D).

7. If the inductor and capacitor have the same reactance, then

$$\frac{1}{\omega C} = \omega L$$
$$\omega = \frac{1}{\sqrt{LC}}$$

The frequency is

$$f = \frac{\omega}{2\pi} = \frac{1}{2\pi\sqrt{LC}}$$
$$= \frac{1}{2\pi\sqrt{(50 \times 10^{-3} \text{ H})(15 \times 10^{-6} \text{ F})}}$$
$$= 184 \text{ Hz} \quad (180 \text{ Hz})$$

The answer is (B).

8. The resonant frequency for a series-RLC circuit is

$$\omega_0 = \frac{1}{\sqrt{LC}} = \frac{1}{\sqrt{(3 \text{ H})(8 \times 10^{-6} \text{ F})}}$$
$$= 204 \text{ rad/s} \quad (200 \text{ rad/s})$$

The answer is (D).

36 Rotating Machines

PRACTICE PROBLEMS

1. A DC generator turns at 2000 rpm and has an output of 200 V. The armature constant is 0.5 V·min/Wb, and the field constant of the machine is 0.02 H. What is most nearly the field current produced by the generator?

(A) 4.0 A
(B) 10 A
(C) 1000 A
(D) 4000 A

2. A two-pole induction motor operates on a three-phase, 60 Hz line-to-line supply. The motor speed is 3420 rpm, and the rms voltage is 240 V. What is most nearly the slip?

(A) 5%
(B) 7%
(C) 10%
(D) 20%

3. A DC generator produces 24 V while operating at 1200 rpm with a magnetic flux of 0.02 Wb. The same generator is operated at 1000 rpm with a magnetic flux of 0.05 Wb. Disregarding armature resistance, what is most nearly the new voltage the generator produces?

(A) 20 V
(B) 50 V
(C) 60 V
(D) 100 V

4. If the slip in a motor decreases while the torque remains constant, what is the effect on the motor's speed and power?

(A) speed increases, power increases
(B) speed increases, power decreases
(C) speed decreases, power increases
(D) speed decreases, power decreases

5. A four-pole induction motor operates on a three-phase, 60 Hz line-to-line supply. The rms voltage is 240 V. The slip is 2%. What is most nearly the operating speed?

(A) 1240 rpm
(B) 1660 rpm
(C) 1760 rpm
(D) 1800 rpm

SOLUTIONS

1. The magnetic flux is

$$V_a = K_a n \phi$$

$$\phi = \frac{V_a}{K_a n} = \frac{200 \text{ V}}{\left(0.5 \; \frac{\text{V·min}}{\text{Wb}}\right)(2000 \text{ rpm})}$$

$$= 0.2 \text{ Wb}$$

Solve for the field current.

$$\phi = K_f I_f$$

$$I_f = \frac{\phi}{K_f} = \frac{0.2 \text{ Wb}}{0.02 \text{ H}}$$

$$= 10 \text{ A}$$

The answer is (B).

2. The synchronous speed is

$$n_s = \frac{120f}{p}$$

$$= \frac{\left(120 \; \frac{\text{rev}}{\text{min} \cdot \text{Hz}}\right)(60 \text{ Hz})}{2}$$

$$= 3600 \text{ rpm}$$

The slip is

$$\text{slip} = \frac{n_s - n}{n_s} \times 100\%$$

$$= \frac{3600 \; \frac{\text{rev}}{\text{min}} - 3420 \; \frac{\text{rev}}{\text{min}}}{3600 \; \frac{\text{rev}}{\text{min}}} \times 100\%$$

$$= 5\%$$

The answer is (A).

3. Since the generated voltage is $V_a = K_a n \phi$, the voltage is proportional to the rotational speed and the flux. The new generated voltage will be

$$V_{\text{new}} = V_{\text{old}} \left(\frac{n_{\text{new}}}{n_{\text{old}}}\right)\left(\frac{\phi_{\text{new}}}{\phi_{\text{old}}}\right)$$

$$= (24 \text{ V}) \left(\frac{1000 \; \frac{\text{rev}}{\text{min}}}{1200 \; \frac{\text{rev}}{\text{min}}}\right)\left(\frac{0.05 \text{ Wb}}{0.02 \text{ Wb}}\right)$$

$$= 50 \text{ V}$$

The answer is (B).

4. Solving the slip equation for the motor speed yields

$$\text{slip} = \frac{n_s - n}{n_s}$$

$$n = n_s(1 - \text{slip})$$

Therefore, the speed of the motor increases as the slip decreases.

Find the relationship between slip and power, P.

$$P = T\Omega = T \left(\frac{60 \; \frac{\text{s}}{\text{min}}}{2\pi \; \frac{\text{rad}}{\text{rev}}}\right) n$$

$$= T \left(\frac{60 \; \frac{\text{s}}{\text{min}}}{2\pi \; \frac{\text{rad}}{\text{rev}}}\right) n_s(1 - \text{slip})$$

Therefore, the power of the motor increases as the slip decreases.

The answer is (A).

5. The synchronous speed is

$$n_s = \frac{120f}{p}$$

$$= \frac{\left(120 \; \frac{\text{rev}}{\text{min} \cdot \text{Hz}}\right)(60 \text{ Hz})}{4}$$

$$= 1800 \text{ rpm}$$

Solving the slip equation for the operating speed yields

$$\text{slip} = \frac{n_s - n}{n_s}$$

$$n = n_s(1 - \text{slip})$$

$$= \left(1800 \; \frac{\text{rev}}{\text{min}}\right)(1 - 0.02)$$

$$= 1764 \text{ rpm} \quad (1760 \text{ rpm})$$

The answer is (C).

37 Kinematics

PRACTICE PROBLEMS

1. A particle's curvilinear motion is represented by the equation $s(t) = 20t + 4t^2 - 3t^3$. What is most nearly the particle's initial velocity?

(A) 20 m/s
(B) 25 m/s
(C) 30 m/s
(D) 32 m/s

2. A vehicle is traveling at 62 km/h when the driver sees a traffic light in an intersection 530 m ahead turn red. The light's red cycle duration is 25 s. The driver wants to enter the intersection without stopping the vehicle, just as the light turns green. If the vehicle decelerates at a constant rate of 0.35 m/s², what will be its approximate speed when the light turns green?

(A) 31 km/h
(B) 43 km/h
(C) 59 km/h
(D) 63 km/h

3. A projectile has an initial velocity of 110 m/s and a launch angle of 20° from the horizontal. The surrounding terrain is level, and air friction is to be disregarded. What is most nearly the flight time of the projectile?

(A) 3.8 s
(B) 7.7 s
(C) 8.9 s
(D) 12 s

4. A particle's position is defined by

$$\mathbf{s}(t) = 2\sin t\,\mathbf{i} + 4\cos t\,\mathbf{j} \quad [t \text{ in radians}]$$

What is most nearly the magnitude of the particle's velocity when $t = 4$ rad?

(A) 2.6
(B) 2.7
(C) 3.3
(D) 4.1

5. A roller coaster train climbs a hill with a constant gradient. During a 10 s period, the acceleration is constant at 0.4 m/s², and the average velocity of the train is 40 km/h. What is most nearly the velocity of the train after 10 s?

(A) 9.1 m/s
(B) 11 m/s
(C) 13 m/s
(D) 15 m/s

6. Choose the equation that best represents a rigid body or particle under constant acceleration.

(A) $a = 9.81 \text{ m/s}^2 + v_0/t$
(B) $v = a_0(t - t_0) + v_0$
(C) $v = v_0 + \int_0^t a(t)\,dt$
(D) $a = v_t^2/r$

7. A particle's curvilinear motion is represented by the equation $s(t) = 40t + 5t^2 - 8t^3$. What is most nearly the initial acceleration of the particle?

(A) 2 m/s²
(B) 3 m/s²
(C) 8 m/s²
(D) 10 m/s²

8. The rotor of a steam turbine is rotating at 7200 rpm when the steam supply is suddenly cut off. The rotor decelerates at a constant rate and comes to rest after 5 min. What is most nearly the angular deceleration of the rotor?

(A) 0.40 rad/s²
(B) 2.5 rad/s²
(C) 5.8 rad/s²
(D) 16 rad/s²

9. The angular position of a car traveling around a curve is described by the following function of time (in seconds).

$$\theta(t) = t^3 - 2t^2 - 4t + 10$$

What is most nearly the angular acceleration of the car at a time of 5 s?

(A) 4.0 rad/s^2

(B) 6.0 rad/s^2

(C) 26 rad/s^2

(D) 30 rad/s^2

10. A vehicle is traveling at 70 km/h when the driver sees a traffic light in the next intersection turn red. The intersection is 250 m away, and the light's red cycle duration is 15 s. What is most nearly the uniform deceleration that will put the vehicle in the intersection the moment the light turns green?

(A) 0.18 m/s^2

(B) 0.25 m/s^2

(C) 0.37 m/s^2

(D) 1.3 m/s^2

11. A projectile has an initial velocity of 85 m/s and a launch angle of 60° from the horizontal. The surrounding terrain is level, and air friction is to be disregarded. What is most nearly the horizontal distance traveled by the projectile?

(A) 80 m

(B) 400 m

(C) 640 m

(D) 1200 m

12. A particle's position is defined by

$$\mathbf{s}(t) = 15 \sin t\mathbf{i} + 8.5 \cos t\mathbf{j} \quad [t \text{ in radians}]$$

What is most nearly the magnitude of the particle's acceleration when $t = \pi$?

(A) 6.5

(B) 8.5

(C) 15

(D) 17

13. A particle's curvilinear motion is represented by the equation $s(t) = 30t - 8t^2 + 6t^3$. What is most nearly the minimum speed reached by the particle?

(A) 26 m/s

(B) 30 m/s

(C) 35 m/s

(D) 48 m/s

14. A projectile has an initial velocity of 80 m/s and a launch angle of 42° from the horizontal. The surrounding terrain is level, and air friction is to be disregarded. What is most nearly the maximum elevation achieved by the projectile?

(A) 72 m

(B) 150 m

(C) 350 m

(D) 620 m

SOLUTIONS

1. The initial velocity at $t=0$ is

$$v = \frac{dr}{dt} = \frac{ds}{dt} = 20 + 8t - 9t^2$$
$$= 20 + (8)(0 \text{ s}) - (9)(0 \text{ s})^2$$
$$= 20 \text{ m/s}$$

The answer is (A).

2. The velocity after 25 s of constant deceleration is

$$v(t) = a_0(t - t_0) + v_0$$
$$= \frac{\left(-0.35 \, \frac{\text{m}}{\text{s}^2}\right)(25 \text{ s} - 0 \text{ s})\left(3600 \, \frac{\text{s}}{\text{h}}\right)}{1000 \, \frac{\text{m}}{\text{km}}} + 62 \, \frac{\text{km}}{\text{h}}$$
$$= 31 \text{ km/h}$$

The answer is (A).

3. The vertical component of velocity is zero at the apex. Calculate the time to reach the apex.

$$v_y = -gt + v_0 \sin(\theta)$$
$$0 = -\left(9.81 \, \frac{\text{m}}{\text{s}^2}\right)t + \left(110 \, \frac{\text{m}}{\text{s}}\right)\sin 20°$$
$$t = 3.84 \text{ s}$$

The projectile takes the same amount of time to return to the ground from the apex as it took to reach the apex after launch. The total flight time is

$$t_{\text{total}} = (2)(3.84 \text{ s}) = 7.67 \text{ s} \quad (7.7 \text{ s})$$

The answer is (B).

4. The velocity is

$$v(t) = \frac{d\mathbf{s}(t)}{dt} = \frac{d}{dt}(2 \sin t \mathbf{i} + 4 \cos t \mathbf{j})$$
$$= 2 \cos t \mathbf{i} - 4 \sin t \mathbf{j}$$

At $t = 4$ rad,

$$\mathbf{v}(4) = 2\cos(4 \text{ rad})\mathbf{i} - 4\sin(4 \text{ rad})\mathbf{j}$$
$$= -1.31\mathbf{i} - (-3.03)\mathbf{j}$$
$$|\mathbf{v}(4)| = \sqrt{(-1.31\mathbf{i})^2 + (3.03\mathbf{i})^2}$$
$$= 3.3$$

The answer is (C).

5. If the train travels for 10 s at an average velocity of 40 km/h, then the distance traveled in 10 s is

$$s(t) = v_{\text{ave}} t = \frac{\left(40 \, \frac{\text{km}}{\text{h}}\right)\left(1000 \, \frac{\text{m}}{\text{km}}\right)(10 \text{ s})}{3600 \, \frac{\text{s}}{\text{h}}}$$
$$= 111.1 \text{ m}$$

Rearrange the equation for distance as a function of initial velocity and acceleration, and solve for the initial velocity.

$$s(t) = a_0(t - t_0)^2/2 + v_0(t - t_0) + s_0$$
$$v_0 = \frac{s(t) - s_0 - \dfrac{a_0(t - t_0)^2}{2}}{t - t_0}$$
$$= \frac{111.1 \text{ m} - 0 \text{ m} - \dfrac{\left(0.4 \, \frac{\text{m}}{\text{s}^2}\right)(10 \text{ s} - 0 \text{ s})^2}{2}}{10 \text{ s} - 0 \text{ s}}$$
$$= 9.11 \text{ m/s}$$

For an initial velocity of 9.11 m/s and an acceleration of 0.4 m/s² over 10 s, the final velocity after 10 s is

$$v_f = v_0 + a_0 t$$
$$= 9.11 \, \frac{\text{m}}{\text{s}} + \left(0.4 \, \frac{\text{m}}{\text{s}^2}\right)(10 \text{ s})$$
$$= 13.11 \text{ m/s} \quad (13 \text{ m/s})$$

The answer is (C).

6. Option A is an expression for acceleration that varies with time. Option C is an expression for velocity with a generalized time-varying acceleration. The expression in option D relates tangential and normal accelerations, respectively, along a curved path, to the tangential velocity. For a generalized curved path, these accelerations are not constant.

Option B is the expression for the velocity of a linear system under constant acceleration.

$$v(t) = a_0 \int dt = a_0(t - t_0) + v_0$$

The answer is (B).

7. The acceleration at $t = 0$ is

$$a = \frac{d^2 r}{dt^2} = \frac{d^2 s}{dt^2} = 10 - 48t$$
$$= 10 \text{ m/s}^2$$

The answer is (D).

8. The angular deceleration (velocity) is

$$\omega_f = \omega_0 - \alpha t$$

$$\alpha = \frac{\omega_0 - \omega_f}{t}$$

$$= \frac{\left(7200 \, \frac{\text{rev}}{\text{min}}\right)\left(2\pi \, \frac{\text{rad}}{\text{rev}}\right) - 0 \, \frac{\text{rad}}{\text{s}}}{(5 \text{ min})\left(60 \, \frac{\text{s}}{\text{min}}\right)^2}$$

$$= 2.51 \text{ rad/s}^2 \quad (2.5 \text{ rad/s}^2)$$

The answer is (B).

9. The angular acceleration is

$$\alpha(t) = \frac{d^2\theta}{dt^2} = 6t - 4$$

$$\alpha(5) = (6)(5 \text{ s}) - 4 = 26 \text{ rad/s}^2$$

The answer is (C).

10. Rearrange the equation for the distance traveled under a constant acceleration. Let the initial distance traveled equal 0 m, and the initial time equal 0 s.

$$s(t) = a_0(t - t_0)^2/2 + v_0(t - t_0) + s_0$$

$$a_0 = \frac{(2)(-v_0(t - t_0) - s(t) - s_0)}{(t - t_0)^2}$$

$$= \frac{(2)\left(\frac{\left(-70 \, \frac{\text{km}}{\text{h}}\right)\left(1000 \, \frac{\text{m}}{\text{km}}\right)(15 \text{ s} - 0 \text{ s})}{3600 \, \frac{\text{s}}{\text{h}}} + 250 \text{ m} - 0 \text{ m}\right)}{(15 \text{ s} - 0 \text{ s})^2}$$

$$= -0.37 \text{ m/s}^2 \quad (0.37 \text{ m/s}^2 \text{ deceleration})$$

The answer is (C).

11. Calculate the total flight time. The vertical component of velocity is zero at the apex.

$$v_y = -gt + v_0 \sin(\theta)$$

$$0 = -\left(9.81 \, \frac{\text{m}}{\text{s}^2}\right)t + \left(85 \, \frac{\text{m}}{\text{s}}\right)\sin 60°$$

$$t = 7.50 \text{ s}$$

The projectile takes the same amount of time to return to the ground from the apex as it took to reach the apex after launch. The total flight time is

$$t = (2)(7.50 \text{ s}) = 15.0 \text{ s}$$

The horizontal distance traveled is

$$x = v_0 \cos(\theta)t + x_0$$

$$= \left(85 \, \frac{\text{m}}{\text{s}}\right)\cos 60°(15.0 \text{ s}) + 0 \text{ m}$$

$$= 638 \text{ m} \quad (640 \text{ m})$$

The answer is (C).

12. The velocity is

$$\mathbf{v}(t) = \frac{d\mathbf{s}(t)}{dt} = \frac{d}{dt}(15 \sin t \mathbf{i} + 8.5 \cos t \mathbf{j})$$

$$= 15 \cos t \mathbf{i} - 8.5 \sin t \mathbf{j}$$

$$\mathbf{a}(t) = \frac{d\mathbf{v}(t)}{dt} = -15 \sin t \mathbf{i} - 8.5 \cos t \mathbf{j}$$

$$\mathbf{a}(\pi) = -15 \sin \pi \mathbf{i} - 8.5 \cos \pi \mathbf{j}$$

$$= 0\mathbf{i} + 8.5\mathbf{j}$$

$$|\mathbf{a}(\pi)| = \sqrt{(0\mathbf{i})^2 + (8.5\mathbf{j})^2}$$

$$= 8.5$$

The answer is (B).

13. The minimum of the velocity function is found by equating the derivative of the velocity function to zero and solving for t.

$$v(t) = \frac{ds}{dt} = \frac{d}{dt}(30t - 8t^2 + 6t^3)$$

$$= 30 - 16t + 18t^2$$

$$\frac{dv}{dt} = \frac{d}{dt}(30 - 16t + 18t^2) = -16 + 36t = 0$$

$$t = 0.444 \text{ s}$$

$$v_{\min} = 30 - 16t + 18t^2$$

$$= 30 - (16)(0.444 \text{ s}) - (18)(0.444 \text{ s})^2$$

$$= 26.4 \text{ m/s} \quad (26 \text{ m/s})$$

The answer is (A).

14. The maximum elevation is achieved when the projectile is at the apex. The vertical component of velocity is zero at the apex. Calculate the time to reach the apex.

$$v_y = -gt + v_0 \sin(\theta)$$

$$0 = -\left(9.81 \, \frac{\text{m}}{\text{s}^2}\right)t + \left(80 \, \frac{\text{m}}{\text{s}}\right)\sin 42°$$

$$t = 5.46 \text{ s}$$

The elevation at time t is

$$y = -gt^2/2 + v_0 \sin(\theta)t + y_0$$

$$= \frac{-\left(9.81 \, \frac{\text{m}}{\text{s}^2}\right)(5.46 \text{ s})^2}{2}$$

$$+ \left(80 \, \frac{\text{m}}{\text{s}}\right)\sin 42°(5.46 \text{ s}) + 0 \text{ m}$$

$$= 146 \text{ m} \quad (150 \text{ m})$$

The answer is (B).

38 Kinetics

PRACTICE PROBLEMS

1. The 52 kg block shown starts from rest at position A and slides down the inclined plane to position B. The coefficient of friction between the block and the plane is $\mu = 0.15$.

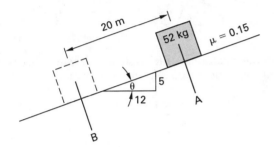

What is most nearly the velocity of the block at position B?

(A) 2.4 m/s

(B) 4.1 m/s

(C) 7.0 m/s

(D) 9.8 m/s

2. A 5 kg block begins from rest and slides down an inclined plane. After 4 s, the block has a velocity of 6 m/s. If the angle of inclination of the plane is 45°, approximately how far has the block traveled after 4 s?

(A) 1.5 m

(B) 3.0 m

(C) 6.0 m

(D) 12 m

3. The elevator in a 20-story apartment building has a mass of 1800 kg. Its maximum velocity and maximum acceleration are 2.5 m/s and 1.4 m/s^2, respectively. A passenger weighing 67 kg stands on a scale in the elevator as the elevator ascends at its maximum acceleration. When the elevator reaches its maximum acceleration, the scale most nearly reads

(A) 67 N

(B) 560 N

(C) 660 N

(D) 750 N

4. A rope is used to tow an 800 kg car with free-rolling wheels over a smooth, level road. The rope will break if the tension exceeds 2000 N. What is most nearly the greatest acceleration that the car can reach without breaking the rope?

(A) 1.2 m/s^2

(B) 2.5 m/s^2

(C) 3.8 m/s^2

(D) 4.5 m/s^2

5. An 8 kg block begins from rest and slides down an inclined plane. After 10 s, the block has a velocity of 15 m/s. The plane's angle of inclination is 30°. What is most nearly the coefficient of friction between the plane and the block?

(A) 0.15

(B) 0.22

(C) 0.40

(D) 0.85

6. If the sum of the forces on a particle is not equal to zero, the particle is

(A) moving with constant velocity in the direction of the resultant force

(B) accelerating in a direction opposite to the resultant force

(C) accelerating in the same direction as the resultant force

(D) moving with a constant velocity opposite to the direction of the resultant force

7. A 383 N horizontal force is applied to the 65 kg block shown. Beginning at position A, the block moves down the slope at a velocity of 12.5 m/s and comes to a complete stop at position B. The coefficient of friction between the block and the plane is $\mu = 0.22$.

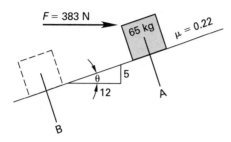

What is most nearly the distance between positions A and B?

(A) 6.1 m
(B) 9.1 m
(C) 15 m
(D) 19 m

SOLUTIONS

1. Choose a coordinate system parallel and perpendicular to the plane, as shown. Let the x-axis be positive in the direction of motion (to the left). Recognize that this is a 5-12-13 triangle. Alternatively, calculate $\sqrt{(5)^2 + (12)^2} = 13$.

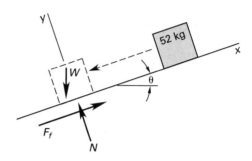

From the equations for friction and the equation for the radial component of force, the acceleration is

$$\sum F_x = ma_x$$
$$W_x - \mu N = ma_x$$
$$mg\sin\theta - \mu mg\cos\theta = ma_x$$
$$a_x = g\sin\theta - \mu g\cos\theta = g(\sin\theta - \mu\cos\theta)$$
$$= \left(9.81\ \tfrac{m}{s^2}\right)\left(\tfrac{5}{13} - (0.15)\left(\tfrac{12}{13}\right)\right)$$
$$= 2.415\ m/s^2$$

The velocity at position B is

$$v^2 = v_0^2 + 2a_x(x - x_0)$$
$$v_0 = x_0 = 0$$
$$v^2 = 2a_x x$$
$$= (2)\left(2.415\ \tfrac{m}{s^2}\right)(20\ m)$$
$$= 96.6\ m^2/s^2$$
$$v = \sqrt{96.6\ \tfrac{m^2}{s^2}} = 9.83\ m/s \quad (9.8\ m/s)$$

The answer is (D).

2. Calculate the initial acceleration.

$$v(t) = a_x t + v_0$$
$$a_x = \frac{v(t) - v_0}{t} = \frac{6\ \tfrac{m}{s} - 0\ \tfrac{m}{s}}{4\ s}$$
$$= 1.5\ m/s^2$$

After 4 s the block will have moved

$$x = \frac{a_x t^2}{2} + v_0 t + x_0$$

$$= \frac{\left(1.5 \ \frac{m}{s^2}\right)(4 \ s)^2}{2} + \left(0 \ \frac{m}{s}\right)(4 \ s) + 0 \ m$$

$$= 12 \ m$$

The answer is (D).

3. Use Newton's second law. The acceleration of the elevator adds to the gravitational acceleration.

$$F = ma = m(a_1 + a_2)$$

$$= (67 \ kg)\left(9.81 \ \frac{m}{s^2} + 1.4 \ \frac{m}{s^2}\right)$$

$$= 751 \ N \quad (750 \ N)$$

The answer is (D).

4. From Newton's second law, the maximum acceleration is

$$F = ma$$

$$a = \frac{F}{m} = \frac{2000 \ N}{800 \ kg}$$

$$= 2.5 \ m/s^2$$

The answer is (B).

5. Calculate the initial acceleration.

$$v_x = a_x t + v_0$$

$$a_x = \frac{v_x - v_0}{t} = \frac{15 \ \frac{m}{s} - 0 \ \frac{m}{s}}{10 \ s}$$

$$= 1.5 \ m/s^2$$

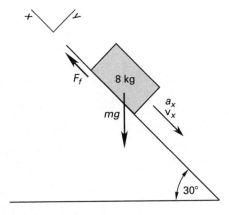

Choose a coordinate system so that the x-direction is parallel to the inclined plane. From the equations for friction, normal force, and the parallel component of force, the coefficient of friction is

$$\sum F_x = ma_x = mg_x - F_f$$

$$ma_x = mg\sin\phi - \mu mg\cos\phi$$

$$\mu = \frac{mg\sin\phi - ma_x}{mg\cos\phi}$$

$$= \frac{g\sin\phi - a_x}{g\cos\phi}$$

$$= \frac{\left(9.81 \ \frac{m}{s^2}\right)\sin 30° - 1.5 \ \frac{m}{s^2}}{\left(9.81 \ \frac{m}{s^2}\right)\cos 30°}$$

$$= 0.40$$

The answer is (C).

6. Newton's second law can be applied separately to any direction in which forces are resolved into components, including the resultant direction.

$$F_R = ma_R$$

Since force and acceleration are both vectors, and mass is a scalar, the direction of acceleration is the same as the resultant force.

$$\mathbf{a}_R = \frac{\mathbf{F}_R}{m}$$

The answer is (C).

7. Choose a coordinate system parallel and perpendicular to the plane, as shown. Let the x-axis be positive in the direction of motion (to the left). Recognize that this is a 5-12-13 triangle. Alternatively, calculate $\sqrt{(5)^2 + (12)^2} = 13$.

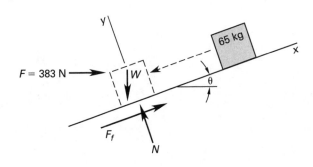

Using the sum of forces and the equations for friction and normal force, solve for the acceleration.

$$\sum F_t = ma_t$$
$$W_x - F_x - \mu N = ma_t$$
$$mg\sin\theta - F\cos\theta - \mu(mg\cos\theta + F\sin\theta) = ma_t$$

$$\begin{aligned}
a_t &= \left(\frac{1}{m}\right)(mg\sin\theta - F\cos\theta - \mu(mg\cos\theta + F\sin\theta)) \\
&= g(\sin\theta - \mu g\cos\theta) - \frac{F}{m}(\cos\theta + \mu\sin\theta) \\
&= \left(9.81\ \frac{\text{m}}{\text{s}^2}\right)\left(\frac{5}{13} - (0.22)\left(\frac{12}{13}\right)\right) \\
&\quad - \left(\frac{383\ \text{N}}{65\ \text{kg}}\right)\left(\frac{12}{13} + (0.22)\left(\frac{5}{13}\right)\right) \\
&= -4.157\ \text{m/s}^2
\end{aligned}$$

The velocity is

$$v^2 = v_0^2 + 2a_t(x - x_0)$$
$$v_0 = 12.5\ \text{m/s}$$
$$v = x_0 = 0$$

The distance between positions A and B is

$$x = \frac{-v_0^2}{2a_t} = \frac{-\left(12.5\ \frac{\text{m}}{\text{s}}\right)^2}{(2)\left(-4.157\ \frac{\text{m}}{\text{s}^2}\right)}$$
$$= 18.79\ \text{m} \quad (19\ \text{m})$$

The answer is (D).

39 Kinetics of Rotational Motion

PRACTICE PROBLEMS

1. A 1530 kg car is towing a 300 kg trailer. The coefficient of friction between all tires and the road is 0.80. The car and trailer are traveling at 100 km/h around a banked curve of radius 200 m. What is most nearly the necessary banking angle such that tire friction will NOT be necessary to prevent skidding?

(A) 8.0°
(B) 21°
(C) 36°
(D) 78°

2. Why does a spinning ice skater's angular velocity increase as she brings her arms in toward her body?

(A) Her mass moment of inertia is reduced.
(B) Her angular momentum is constant.
(C) Her radius of gyration is reduced.
(D) all of the above

3. A 1 m long uniform rod has a mass of 10 kg. It is pinned at one end to a frictionless pivot. What is most nearly the mass moment of inertia of the rod taken about the pivot point?

(A) 0.83 kg·m^2
(B) 2.5 kg·m^2
(C) 3.3 kg·m^2
(D) 10 kg·m^2

4. In the linkage mechanism shown, link AB rotates with an instantaneous counterclockwise angular velocity of 10 rad/s.

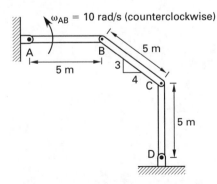

What is most nearly the instantaneous angular velocity of link BC when link AB is horizontal and link CD is vertical?

(A) 2.3 rad/s (clockwise)
(B) 3.3 rad/s (counterclockwise)
(C) 5.5 rad/s (clockwise)
(D) 13 rad/s (clockwise)

5. Two 2 kg blocks are linked as shown.

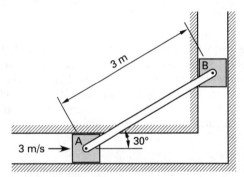

Assuming that the surfaces are frictionless, what is most nearly the velocity of block B if block A is moving at a speed of 3 m/s?

(A) 0 m/s
(B) 1.3 m/s
(C) 1.7 m/s
(D) 5.2 m/s

6. A car travels on a perfectly horizontal, unbanked circular track of radius r. The coefficient of friction between the tires and the track is 0.3. If the car's velocity is 10 m/s, what is most nearly the smallest radius the car can travel without skidding?

(A) 10 m
(B) 34 m
(C) 50 m
(D) 68 m

7. A uniform rod (AB) of length L and weight W is pinned at point C. The rod starts from rest and accelerates with an angular acceleration of $12g/7L$.

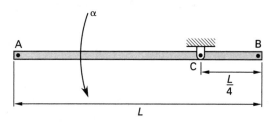

What is the instantaneous reaction at point C at the moment rotation begins?

(A) $\dfrac{W}{4}$

(B) $\dfrac{W}{3}$

(C) $\dfrac{4W}{7}$

(D) $\dfrac{7W}{12}$

8. A wheel with a 0.75 m radius has a mass of 200 kg. The wheel is pinned at its center and has a radius of gyration of 0.25 m. A rope is wrapped around the wheel and supports a hanging 100 kg block. When the wheel is released, the rope begins to unwind. What is most nearly the angular acceleration of the wheel as the block descends?

(A) 5.9 rad/s²

(B) 6.5 rad/s²

(C) 11 rad/s²

(D) 14 rad/s²

9. A car travels around an unbanked 50 m radius curve without skidding. The coefficient of friction between the tires and road is 0.3. What is most nearly the car's maximum velocity?

(A) 14 km/h

(B) 25 km/h

(C) 44 km/h

(D) 54 km/h

10. A uniform rod (AB) of length L and weight W is pinned at point C and restrained by cable OA. The cable is suddenly cut. The rod starts to rotate about point C, with point A moving down and point B moving up.

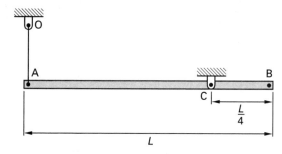

The instantaneous linear acceleration of point B is

(A) $\dfrac{3g}{16}$

(B) $\dfrac{g}{4}$

(C) $\dfrac{3g}{7}$

(D) $\dfrac{3g}{4}$

SOLUTIONS

1. The necessary superelevation angle without relying on friction is

$$\theta = \arctan \frac{v^2}{gr}$$

$$= \arctan \frac{\left(\left(100 \ \frac{km}{h}\right)\left(1000 \ \frac{m}{km}\right)\right)^2}{\left(9.81 \ \frac{m}{s^2}\right)(200 \ m)\left(3600 \ \frac{s}{h}\right)^2}$$

$$= 21.47° \quad (21°)$$

The answer is (B).

2. As the skater brings her arms in, her radius of gyration and mass moment of inertia decrease. However, in the absence of friction, her angular momentum, H, is constant.

$$\omega = \frac{H}{I}$$

Since angular velocity, ω, is inversely proportional to the mass moment of inertia, the angular velocity increases when the mass moment of inertia decreases.

The answer is (D).

3. The mass moment of inertia of the rod taken about one end is

$$I_{rod} = \frac{ML^2}{3} = \frac{(10 \ kg)(1 \ m)^2}{3}$$

$$= 3.33 \ kg \cdot m^2 \quad (3.3 \ kg \cdot m^2)$$

The answer is (C).

4. Find the instantaneous center of rotation. The absolute velocity directions at points B and C are known. The instantaneous center is located by drawing perpendiculars to these velocities, as shown. The angular velocity of any point on rigid body link BC is the same at this instant.

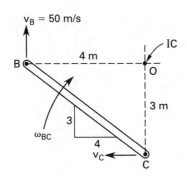

The velocity of point B is

$$v_B = AB\omega_{AB} = (5 \ m)\left(10 \ \frac{rad}{s}\right) = 50 \ m/s$$

The angular velocity of link BC is

$$\omega_{BC} = \frac{v_B}{OB} = \frac{50 \ \frac{m}{s}}{4 \ m}$$

$$= 12.5 \ rad/s \quad (13 \ rad/s) \quad [clockwise]$$

The answer is (D).

5. The instantaneous center of rotation for the slider rod assembly can be found by extending perpendiculars from the velocity vectors, as shown. Both blocks can be assumed to rotate about point C with angular velocity ω.

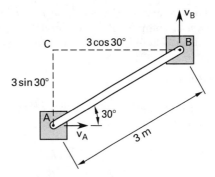

The velocity of block B is

$$\omega = \frac{v_A}{CA} = \frac{v_B}{BC}$$

$$v_B = \frac{v_A BC}{CA} = \frac{\left(3 \ \frac{m}{s}\right)(3 \ m)\cos 30°}{(3 \ m)\sin 30°}$$

$$= 5.2 \ m/s$$

The answer is (D).

6. If there is no skidding, the frictional force, F_f, will equal the centrifugal force, F_c. From the equations for centrifugal force and frictional force, the smallest possible radius is

$$F_c = \frac{mv^2}{r}$$

$$F_f = \mu N = \mu mg$$

$$\frac{mv^2}{r} = \mu mg$$

$$r = \frac{v^2}{\mu g} = \frac{\left(10 \ \frac{m}{s}\right)^2}{(0.3)\left(9.81 \ \frac{m}{s^2}\right)} = 34 \ m$$

The answer is (B).

7. The mass moment of inertia of the rod about its center of gravity is

$$I_{CG} = \frac{ML^2}{12} = \left(\frac{W}{g}\right)\left(\frac{L^2}{12}\right)$$

Take moments about the center of gravity of the rod. All moments due to gravitational forces will cancel. The only unbalanced force acting on the rod will be the vertical reaction force, R_C, at point C.

$$\sum M_{CG} = R_C\left(\frac{L}{4}\right) = I_{CG}\alpha_{CG}$$

$$R_C\left(\frac{L}{4}\right) = \left(\left(\frac{W}{g}\right)\left(\frac{L^2}{12}\right)\right)\left(\frac{12g}{7L}\right)$$

$$R_C = \frac{4W}{7}$$

The angular velocity is zero, so the center of the mass does not have a component of acceleration in the horizontal direction. There is no horizontal force component at point C.

The answer is (C).

8. From the equation for the radius of gyration, the mass moment of inertia of the wheel is

$$r = \sqrt{\frac{I}{m_{wheel}}}$$

$$I = m_{wheel}r^2$$

The unbalanced moment on the wheel is

$$M = FR = (mg - ma)R = mR(g - a)$$
$$= m_{block}R(g - R\alpha)$$

The acceleration is given by

$$M = I\alpha$$
$$m_{block}R(g - R\alpha) = m_{wheel}r^2\alpha$$

Combine the equations and solve.

$$\alpha = \frac{m_{block}Rg}{m_{wheel}r^2 + m_{block}R^2}$$

$$= \frac{(100 \text{ kg})(0.75 \text{ m})\left(9.81 \frac{\text{m}}{\text{s}^2}\right)}{(200 \text{ kg})(0.25 \text{ m})^2 + (100 \text{ kg})(0.75 \text{ m})^2}$$

$$= 10.7 \text{ rad/s}^2 \quad (11 \text{ rad/s}^2)$$

The answer is (C).

9. If the car does not skid, the frictional force and the centrifugal force must be equal. From the equations for centrifugal force and frictional force, the car's maximum velocity is

$$F_c = \frac{mv^2}{r}$$

$$F_f = \mu N = \mu mg$$

$$\frac{mv^2}{r} = \mu mg$$

$$v = \sqrt{\mu gr}$$

$$= \sqrt{(0.3)\left(\frac{\left(9.81 \frac{\text{m}}{\text{s}^2}\right)\left(60 \frac{\text{s}}{\text{min}}\right)^2\left(60 \frac{\text{min}}{\text{h}}\right)^2}{1000 \frac{\text{m}}{\text{km}}}\right) \times \left(\frac{50 \text{ m}}{1000 \frac{\text{m}}{\text{km}}}\right)}$$

$$= 43.67 \text{ km/h} \quad (44 \text{ km/h})$$

The answer is (C).

10. Point C is $L/4$ from the center of gravity of the rod. The mass moment of inertia about point C is

$$I_C = I_{CG} + Md^2 = \frac{ML^2}{12} + M\left(\frac{L}{4}\right)^2 = \left(\frac{7}{48}\right)ML^2$$

The sum of moments on the rod is

$$\sum M_C = \sum Fr = \left(\frac{3W}{4}\right)\left(\frac{\frac{3L}{4}}{2}\right) - \left(\frac{W}{4}\right)\left(\frac{\frac{L}{4}}{2}\right)$$

$$= \frac{WL}{4}$$

$$= \frac{MgL}{4}$$

The angular acceleration is

$$\alpha = \frac{\sum M_C}{I_C} = \frac{\frac{MgL}{4}}{\left(\frac{7}{48}\right)ML^2} = \frac{12g}{7L}$$

The tangential acceleration of point B is

$$a_{t,B} = r\alpha = \left(\frac{L}{4}\right)\left(\frac{12g}{7L}\right) = \frac{3g}{7}$$

The answer is (C).

40 Energy and Work

PRACTICE PROBLEMS

1. The 40 kg mass, m, in the illustration shown is guided by a frictionless rail. The spring constant, k, is 3000 N/m. The spring is compressed sufficiently and released, such that the mass barely reaches point A.

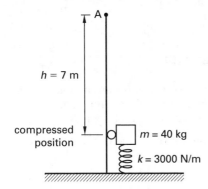

What is most nearly the initial spring compression?

(A) 0.96 m

(B) 1.3 m

(C) 1.4 m

(D) 1.8 m

2. Two balls, both of mass 2 kg, collide head on. The velocity of each ball at the time of the collision is 2 m/s. The coefficient of restitution is 0.5. Most nearly, what are the final velocities of the balls?

(A) 1 m/s and −1 m/s

(B) 2 m/s and −2 m/s

(C) 3 m/s and −3 m/s

(D) 4 m/s and −4 m/s

3. A 1500 kg car traveling at 100 km/h is towing a 250 kg trailer. The coefficient of friction between the tires and the road is 0.8 for both the car and trailer. Approximately what energy is dissipated by the brakes if the car and trailer are braked to a complete stop?

(A) 96 kJ

(B) 390 kJ

(C) 580 kJ

(D) 680 kJ

4. A 3500 kg car traveling at 65 km/h skids and hits a wall 3 s later. The coefficient of friction between the tires and the road is 0.60. What is most nearly the speed of the car when it hits the wall?

(A) 0.14 m/s

(B) 0.40 m/s

(C) 5.1 m/s

(D) 6.2 m/s

5. In the illustration shown, the 170 kg mass, m, is guided by a frictionless rail. The spring is compressed sufficiently and released, such that the mass barely reaches point B.

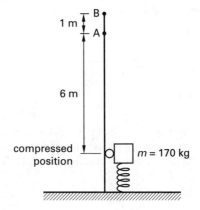

What is most nearly the kinetic energy of the mass at point A?

(A) 20 J

(B) 220 J

(C) 390 J

(D) 1700 J

6. A pickup truck is traveling forward at 25 m/s. The bed is loaded with boxes whose coefficient of friction with the bed is 0.40. What is most nearly the shortest time that the truck can be brought to a stop such that the boxes do not shift?

(A) 2.3 s

(B) 4.7 s

(C) 5.9 s

(D) 6.4 s

7. Two balls both have a mass of 8 kg and collide head on. The velocity of each ball at the time of collision is 18 m/s. The velocity of each ball decreases to 10 m/s in opposite directions after the collision. Approximately how much energy is lost in the collision?

(A) 0.57 kJ
(B) 0.91 kJ
(C) 1.8 kJ
(D) 2.3 kJ

8. The impulse-momentum principle is mostly useful for solving problems involving

(A) force, velocity, and time
(B) force, acceleration, and time
(C) velocity, acceleration, and time
(D) force, velocity, and acceleration

9. A 12 kg aluminum box is dropped from rest onto a large wooden beam. The box travels 0.2 m before contacting the beam. After impact, the box bounces 0.05 m above the beam's surface. Approximately what impulse does the beam impart on the box?

(A) 8.6 N·s
(B) 12 N·s
(C) 36 N·s
(D) 42 N·s

10. The 85 kg mass, m, shown is guided by a frictionless rail. The spring is compressed sufficiently and released, such that the mass barely reaches point B. The spring constant, k, is 1500 N/m.

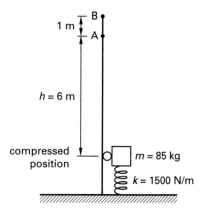

What is most nearly the velocity of the mass at point A?

(A) 3.1 m/s
(B) 4.4 m/s
(C) 9.8 m/s
(D) 20 m/s

SOLUTIONS

1. At the point just before the spring is released, all of the energy in the system is elastic potential energy; while at point A, all of the energy is potential energy due to gravity.

$$\frac{kx^2}{2} = mgh$$

$$x = \sqrt{\frac{2mgh}{k}}$$

$$= \sqrt{\frac{(2)(40 \text{ kg})\left(9.81 \, \frac{\text{m}}{\text{s}^2}\right)(7 \text{ m})}{3000 \, \frac{\text{N}}{\text{m}}}}$$

$$= 1.35 \text{ m} \quad (1.4 \text{ m})$$

The answer is (C).

2. Since the two velocities are in opposite directions, let the velocity of one ball, v_1, equal 2 m/s and the velocity of the other ball, v_2, equal -2 m/s.

From the definition of the coefficient of restitution,

$$e = \frac{v_2' - v_1'}{v_1 - v_2}$$

$$v_2' - v_1' = e(v_1 - v_2)$$

$$= (0.5)\left(2 \, \frac{\text{m}}{\text{s}} - \left(-2 \, \frac{\text{m}}{\text{s}}\right)\right)$$

$$= 2 \text{ m/s} \qquad [\text{Eq. I}]$$

From the conservation of momentum,

$$m_1 v_1 + m_2 v_2 = m_1 v_1' + m_2 v_2'$$

But, $m_1 = m_2$.

$$v_1 + v_2 = v_1' + v_2'$$

Since $v_1 = 2$ m/s and $v_2 = -2$ m/s,

$$v_1 + v_2 = 2 \, \frac{\text{m}}{\text{s}} + \left(-2 \, \frac{\text{m}}{\text{s}}\right) = 0$$

So,

$$v_1' + v_2' = 0 \qquad [\text{Eq. II}]$$

Solve Eq. I and Eq. II simultaneously by adding them.

$$v_1' = -1 \text{ m/s}$$
$$v_2' = 1 \text{ m/s}$$

The answer is (A).

3. The original velocity of the car and trailer is

$$v = \frac{\left(100 \, \frac{\text{km}}{\text{h}}\right)\left(1000 \, \frac{\text{m}}{\text{km}}\right)}{\left(60 \, \frac{\text{s}}{\text{min}}\right)\left(60 \, \frac{\text{min}}{\text{h}}\right)} = 27.78 \text{ m/s}$$

Since the final velocity is zero, the energy dissipated is the original kinetic energy.

$$T = \frac{m\text{v}^2}{2} = \frac{(1500 \text{ kg} + 250 \text{ kg})\left(27.78 \, \frac{\text{m}}{\text{s}}\right)^2}{2}$$
$$= 675\,154 \text{ J} \quad (680 \text{ kJ})$$

The answer is (D).

4. The frictional force (negative because it opposes motion) decelerating the car is

$$F_f = -\mu N = -\mu mg$$
$$= -(0.60)(3500 \text{ kg})\left(9.81 \, \frac{\text{m}}{\text{s}^2}\right)$$
$$= -20\,601 \text{ N}$$

Use the impulse-momentum principle.

$$F_f(t_1 - t_2) = m(\text{v}_1 - \text{v}_2)$$
$$\text{v}_2 = \text{v}_1 - \frac{F_f(t_1 - t_2)}{m}$$
$$= \frac{\left(65 \, \frac{\text{km}}{\text{h}}\right)\left(1000 \, \frac{\text{m}}{\text{km}}\right)}{\left(60 \, \frac{\text{s}}{\text{min}}\right)\left(60 \, \frac{\text{min}}{\text{h}}\right)}$$
$$- \frac{(-20\,601 \text{ N})(0 \text{ s} - 3 \text{ s})}{3500 \text{ kg}}$$
$$= 0.3976 \text{ m/s} \quad (0.40 \text{ m/s})$$

The answer is (B).

5. At point A, the energy of the mass is a combination of kinetic and gravitational potential energies. The total energy of the system is constant, and the kinetic energy at B is 0.

$$E_A = E_B$$
$$U_A + T_A = U_B$$
$$mgh + \frac{m\text{v}^2}{2} = mg(h + 1 \text{ m})$$
$$T_A = mg(h + 1 \text{ m}) - mgh$$
$$= mg(1 \text{ m})$$
$$= (170 \text{ kg})\left(9.81 \, \frac{\text{m}}{\text{s}^2}\right)(1 \text{ m})$$
$$= 1670 \text{ J} \quad (1700 \text{ J})$$

The answer is (D).

6. The frictional force is the only force preventing the boxes from shifting. The forces on each box are its weight, the normal force, and the frictional force. The normal force on each box is equal to the box weight.

$$N = W = mg$$

The frictional force is

$$F_f = \mu N = \mu mg$$

Use the impulse-momentum principle. $\text{v}_2 = 0$. The frictional force is opposite of the direction of motion, so it is negative.

$$\text{Imp} = \Delta p$$
$$F_f \Delta t = m \Delta \text{v}$$
$$\Delta t = \frac{m(\text{v}_2 - \text{v}_1)}{F_f} = \frac{-m\text{v}_1}{-\mu mg} = \frac{\text{v}_1}{\mu g}$$
$$= \frac{25 \, \frac{\text{m}}{\text{s}}}{(0.40)\left(9.81 \, \frac{\text{m}}{\text{s}^2}\right)}$$
$$= 6.37 \text{ s} \quad (6.4 \text{ s})$$

The answer is (D).

7. Each ball possesses kinetic energy before and after the collision. The velocity of each ball is reduced from $|18 \text{ m/s}|$ to $|10 \text{ m/s}|$.

$$\Delta T = T_2 - T_1 = (2)\left(\frac{m(\text{v}_2^2 - \text{v}_1^2)}{2}\right)$$
$$= (2)\left(\frac{(8 \text{ kg})\left(\left(18 \, \frac{\text{m}}{\text{s}}\right)^2 - \left(10 \, \frac{\text{m}}{\text{s}}\right)^2\right)}{2}\right)$$
$$= 1792 \text{ J} \quad (1.8 \text{ kJ})$$

The answer is (C).

8. Impulse is calculated from force and time. Momentum is calculated from mass and velocity. The impulse-momentum principle is useful in solving problems involving force, time, velocity, and mass.

The answer is (A).

9. Initially, the box has potential energy only. (This takes the beam's upper surface as the reference plane.) When the box reaches the beam, all of the potential energy will have been converted to kinetic energy.

$$mgh_1 = \frac{mv_1^2}{2}$$

$$\begin{aligned}v_1 &= \sqrt{2gh_1} \\ &= \sqrt{(2)\left(9.81\ \tfrac{m}{s^2}\right)(0.2\ m)} \\ &= 1.98\ m/s\quad [\text{downward}]\end{aligned}$$

When the box rebounds to its highest point, all of its remaining energy will be potential energy once again.

$$mgh_2 = \frac{mv_2^2}{2}$$

$$\begin{aligned}v_2 &= \sqrt{2gh_2} \\ &= \sqrt{(2)\left(9.81\ \tfrac{m}{s^2}\right)(0.05\ m)} \\ &= 0.99\ m/s\quad [\text{upward}]\end{aligned}$$

Use the impulse-momentum principle. (Downward is taken as the positive velocity direction.)

$$\begin{aligned}\text{Imp} = \Delta p &= m(v_1 - v_2) \\ &= (12\ kg)\left(1.98\ \tfrac{m}{s} - \left(-0.99\ \tfrac{m}{s}\right)\right) \\ &= 35.66\ N{\cdot}s\quad (36\ N{\cdot}s)\end{aligned}$$

The answer is (C).

10. At point A, the energy of the mass is a combination of kinetic and gravitational potential energies. The total energy of the system is constant, and the kinetic energy at B is 0.

$$E_A = E_B$$
$$U_A + T_A = U_B$$
$$mgh + \frac{mv^2}{2} = mg(h + 1\ m)$$

$$\begin{aligned}T_A &= mg(h + 1\ m) - mgh \\ &= mg(1\ m) \\ &= (85\ kg)\left(9.81\ \tfrac{m}{s^2}\right)(1\ m) \\ &= 833.9\ J\end{aligned}$$

Therefore, the velocity of the mass at point A is

$$T_A = \frac{mv^2}{2} = 833.9\ J$$

$$\begin{aligned}v &= \sqrt{\frac{2T_A}{m}} \\ &= \sqrt{\frac{(2)(833.9\ J)}{85\ kg}} \\ &= 4.43\ m/s\quad (4.4\ m/s)\end{aligned}$$

The answer is (B).

41 Vibrations

PRACTICE PROBLEMS

1. A 54 kg mass is supported by three springs as shown. The initial displacement is 5.0 cm downward from the static equilibrium position. No external forces act on the mass after it is released.

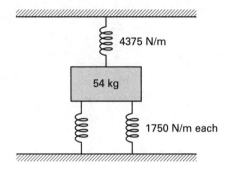

What is most nearly the equivalent spring constant?

(A) 6200 N/m
(B) 6800 N/m
(C) 7400 N/m
(D) 7900 N/m

2. A torsional pendulum is composed of a 0.5 m long shaft with a diameter of 0.5 cm and a shear modulus of 8×10^4 MPa, and a thin 3 kg disk with a radius of 30 cm. The natural frequency of the pendulum is 8.5 rad/s. What is most nearly the period of oscillation?

(A) 0.61 s
(B) 0.74 s
(C) 0.87 s
(D) 0.96 s

3. A pump with a mass of 45 kg is supported by four springs, each having a spring constant of 1750 N/m. The motor is constrained to allow only vertical movement. The natural frequency of the pump is most nearly

(A) 6.0 rad/s
(B) 9.0 rad/s
(C) 12 rad/s
(D) 15 rad/s

4. A massless sheave is supported by two identical springs connected through a frictionless band around it. A mass is carried by a connection to the sheave's center.

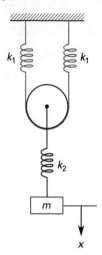

The natural frequency of the system is designed to be 10 rad/s. The spring constant k_2 is half of spring constant k_1, and the mass, m, is 1 kg. The mass associated with the other components is negligible. For the given natural frequency, the spring constant k_1 is most nearly

(A) 3.0 N/m
(B) 40 N/m
(C) 120 N/m
(D) 250 N/m

5. A 68 kg mass is supported by three springs as shown. The initial displacement is 15 cm downward from the static equilibrium position. No external forces act on the mass after it is released. The equivalent spring constant is 5600 N/m.

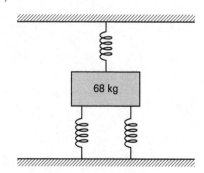

What is most nearly the static deflection of the mass?

(A) 0.20 m

(B) 0.25 m

(C) 0.30 m

(D) 0.40 m

6. A cantilever beam deflects 5 cm when a force of 5 kN is applied at the end. The beam tip is subsequently supported by a spring with a stiffness of 1.5 kN/cm.

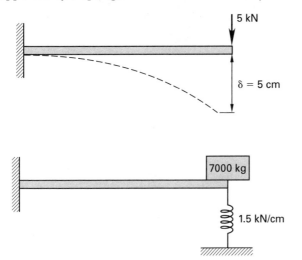

When a 7000 kg mass is attached at the beam's tip, what is most nearly the natural frequency of the mass-beam-spring system?

(A) 1.5 rad/s

(B) 3.1 rad/s

(C) 6.0 rad/s

(D) 6.3 rad/s

7. What is most nearly the natural frequency, ω, of an oscillating body whose period of oscillation is 1.8 s?

(A) 1.8 rad/s

(B) 2.7 rad/s

(C) 3.5 rad/s

(D) 4.2 rad/s

8. What is most nearly the period of a pendulum that passes the center point 20 times a minute?

(A) 0.2 s

(B) 0.3 s

(C) 3 s

(D) 6 s

9. A 28 kg mass is supported by three springs as shown. The initial displacement is 2 cm downward from the static equilibrium position. The static deflection of the mass is 0.21 m, and the equivalent spring constant is 1300 N/m. No external forces act on the mass after it is released.

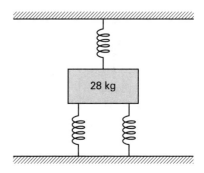

What is most nearly the natural frequency?

(A) 4.0 rad/s

(B) 6.8 rad/s

(C) 12 rad/s

(D) 16 rad/s

SOLUTIONS

1. Since the springs are in parallel, they all share the applied load. The equivalent spring constant is

$$k_{eq} = k_1 + k_2 + k_3 = 4375 \ \frac{N}{m} + 1750 \ \frac{N}{m} + 1750 \ \frac{N}{m}$$
$$= 7875 \ N/m \quad (7900 \ N/m)$$

The answer is (D).

2. The period of oscillation is

$$\tau = \frac{2\pi}{\omega} = \frac{2\pi}{8.5 \ \frac{rad}{s}} = 0.737 \ s \quad (0.74 \ s)$$

The answer is (B).

3. Calculate the total spring constant.

$$k = (4)\left(1750 \ \frac{N}{m}\right) = 7000 \ N/m$$

Calculate the static deflection.

$$mg = k\delta_{st}$$
$$\delta_{st} = \frac{mg}{k} = \frac{(45 \ kg)\left(9.81 \ \frac{m}{s^2}\right)}{7000 \ \frac{N}{m}}$$
$$= 0.063 \ m$$

Calculate the natural frequency.

$$\omega = \sqrt{\frac{g}{\delta_{st}}} = \sqrt{\frac{9.81 \ \frac{m}{s^2}}{0.063 \ m}}$$
$$= 12.48 \ rad/s \quad (12 \ rad/s)$$

The answer is (C).

4. The equivalent system is represented by

$$\ddot{x} + \left(\frac{k}{m}\right)x = 0$$

$$\omega = \sqrt{\frac{k}{m}} \quad \text{[Eq. I]}$$

$$k = \left(\frac{1}{2k_1} + \frac{1}{k_2}\right)^{-1} = \left(\frac{k_2 + 2k_1}{2k_1 k_2}\right)^{-1} \quad \text{[Eq. II]}$$

For $k_2 = \frac{1}{2}k_1$, $m = 1$ kg, $\omega = 10$ rad/s, and $k = \frac{2}{5}k_1$, using Eq. I and Eq. II,

$$\omega = \sqrt{\frac{k}{m}}$$
$$10 \ \frac{rad}{s} = \sqrt{\frac{\frac{2}{5}k_1}{1 \ kg}}$$
$$k_1 = 250 \ N/m$$

The answer is (D).

5. The static deflection is

$$mg = k\delta_{st}$$
$$\delta_{st} = \frac{mg}{k} = \frac{(68 \ kg)\left(9.81 \ \frac{m}{s^2}\right)}{5600 \ \frac{N}{m}}$$
$$= 0.119 \ m \quad (0.20 \ m)$$

The answer is (A).

6. A spring-propped cantilever with an end mass m can be modeled as follows.

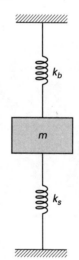

The spring constant k_b is

$$k_b = \frac{m}{\delta_{st}} = \frac{(5 \ kN)\left(1000 \ \frac{N}{kN}\right)}{5 \ cm} = 1000 \ N/cm$$

Both springs undergo the same deflection. The total spring constant is

$$k = k_b + k_s = 1000 \ \frac{\text{N}}{\text{cm}} + \left(1.5 \ \frac{\text{kN}}{\text{cm}}\right)\left(1000 \ \frac{\text{N}}{\text{kN}}\right)$$
$$= 2500 \ \text{N/cm}$$

The natural frequency is

$$\omega = \sqrt{\frac{k}{m}} = \sqrt{\frac{\left(2500 \ \frac{\text{N}}{\text{cm}}\right)\left(100 \ \frac{\text{cm}}{\text{m}}\right)}{7000 \ \text{kg}}}$$
$$= 5.98 \ \text{rad/s} \quad (6.0 \ \text{rad/s})$$

The answer is (C).

7. The natural frequency is

$$\tau = \frac{2\pi}{\omega_n}$$
$$\omega_n = \frac{2\pi}{\tau} = \frac{2\pi}{1.8 \ \text{s}}$$
$$= 3.5 \ \text{rad/s}$$

The answer is (C).

8. A pendulum will pass the center point two times during each complete cycle. Therefore, 10 cycles are completed in 60 s.

$$T = \frac{\text{elapsed time}}{\text{no. of cycles}} = \frac{60 \ \text{s}}{10}$$
$$= 6 \ \text{s}$$

The answer is (D).

9. The natural frequency is

$$\omega = \sqrt{\frac{g}{\delta_{st}}} = \sqrt{\frac{9.81 \ \frac{\text{m}}{\text{s}^2}}{0.21 \ \text{m}}}$$
$$= 6.83 \ \text{rad/s} \quad (6.8 \ \text{rad/s})$$

The answer is (B).

42 Fasteners

PRACTICE PROBLEMS

1. A bolted joint with a joint coefficient of 0.5 experiences an alternating external tensile load between 0 kN and 10 kN. The bolt is initially preloaded to 20 kN. What is most nearly the maximum bolt load?

(A) 15 kN
(B) 20 kN
(C) 25 kN
(D) 30 kN

2. An M12 × 1.75 bolt has an area of 113 mm² and a tensile stress area of 85 mm². The unthreaded shank length and the length of threaded section within the grip are both 12 mm. The modulus of elasticity is 200 GPa. What is most nearly the stiffness of the bolt?

(A) 540 kN/mm
(B) 680 kN/mm
(C) 750 kN/mm
(D) 810 kN/mm

3. What is most nearly the externally applied load on a bolt at joint separation if the bolt preload is 200 kN and the joint coefficient is 0.2?

(A) 40 kN
(B) 200 kN
(C) 300 kN
(D) 1000 kN

SOLUTIONS

1. The maximum bolt load is
$$F_{b,\max} = CP + F_i = (0.5)(10 \text{ kN}) + 20 \text{ kN}$$
$$= 25 \text{ kN}$$

The answer is (C).

2. The stiffness of the bolt is
$$k_b = \frac{A_d A_t E}{A_d l_t + A_t l_d}$$
$$= \frac{(113 \text{ mm}^2)(85 \text{ mm}^2)(200 \text{ GPa})\left(10^6 \frac{\text{kPa}}{\text{GPa}}\right)}{\left(\begin{array}{c}(113 \text{ mm}^2)(12 \text{ mm}) \\ + (85 \text{ mm}^2)(12 \text{ mm})\end{array}\right)\left(1000 \frac{\text{mm}}{\text{m}}\right)^2}$$
$$= 809 \text{ kN/mm} \quad (810 \text{ kN/mm})$$

The answer is (D).

3. At joint separation, the members are no longer compressed, and the member load, F_m, is 0. Set the member load equal to 0, and rearrange to solve for the externally applied load, P.

$$F_m = (1-C)P - F_i = 0$$
$$P = \frac{F_i}{1-C}$$
$$= \frac{200 \text{ kN}}{1-0.2}$$
$$= 250 \text{ kN} \quad (300 \text{ kN})$$

The answer is (C).

43 Machine Design

PRACTICE PROBLEMS

1. A simple epicyclic gearbox with one planet has gears with 24, 40, and 104 teeth on the sun, planet, and internal ring gears, respectively. The sun rotates clockwise at 50 rpm. The ring gear is fixed. The rotational velocity of the planet carrier is most nearly

(A) 9.4 rpm
(B) 12 rpm
(C) 17 rpm
(D) 23 rpm

2. A pressure control valve for limiting hydraulic system pressure is redesigned to use a helical spring with squared and ground ends. The spring will be made from 2.5 mm diameter wire having a shear modulus of 83 GPa. The maximum system operating pressure is to be 1.4 MPa. The effective control area where the spring ultimately acts to limit system pressure is 80 mm^2. The maximum spring deflection is to be 11 mm. The spring index is 8. The spring must have

(A) 2 coils
(B) 4 coils
(C) 6 coils
(D) 7 coils

3. Refer to the epicyclic gear set illustrated. Gear A rotates counterclockwise on a fixed center at 100 rpm. The ring gear rotates. The planet carrier rotates clockwise at 60 rpm. Each gear has the number of teeth indicated.

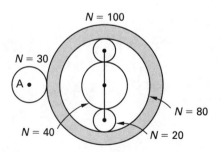

What is most nearly the rotational speed of the sun gear?

(A) 30 rpm
(B) 60 rpm
(C) 90 rpm
(D) 120 rpm

4. An 18 tooth straight spur gear transmits a torque of 1100 N·mm. The pitch circle diameter is 19 mm, and the pressure angle is 14.5°. What is most nearly the radial force on the gear?

(A) 16 N
(B) 30 N
(C) 110 N
(D) 120 N

5. Two concentric springs are constructed with squared and ground ends from oil-hardened steel. The spring dimensions and properties are as follows.

inner spring	
wire diameter	4.5 mm
mean coil diameter	38 mm
total number of coils	12.75
outer spring	
wire diameter	5.723 mm
mean coil diameter	51 mm
total number of coils	10.25

How many active coils are in the inner and outer spring, respectively?

(A) $N_{\text{inner}} = 8$; $N_{\text{outer}} = 11$
(B) $N_{\text{inner}} = 11$; $N_{\text{outer}} = 9$
(C) $N_{\text{inner}} = 12$; $N_{\text{outer}} = 9$
(D) $N_{\text{inner}} = 15$; $N_{\text{outer}} = 12$

6. A spring with 12 active coils and a spring index of 9 supports a static load of 220 N with a deflection of 12 mm. The shear modulus of the spring material is 83 GPa. What is most nearly the theoretical wire diameter?

(A) 15 mm
(B) 17 mm
(C) 18 mm
(D) 19 mm

7. A simple epicyclic gear set has two planets. The sun gear has 32 teeth, the planets have 16 teeth, and the ring has 64 teeth. The sun gear turns clockwise at 100 rpm. The ring is fixed. What is most nearly the rotational speed of the carrier?

(A) 20 rpm
(B) 25 rpm
(C) 33 rpm
(D) 50 rpm

8. For the four-bar linkage shown, at what approximate angle β does $\omega_2 = \omega_4$?

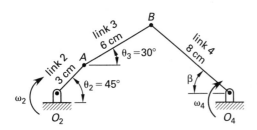

(A) 50°
(B) 100°
(C) 140°
(D) 160°

9. For the four-bar linkage shown, what is most nearly the angular velocity of link 4 if $\omega_2 = 2$ rad/s (clockwise rotation) and clockwise rotation is positive?

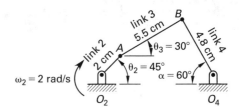

(A) −0.43 rad/s
(B) −0.22 rad/s
(C) 0.41 rad/s
(D) 0.77 rad/s

10. Two concentric springs are constructed with plain ends from oil-hardened steel. The spring dimensions and properties are as follows.

inner spring	
wire diameter	6.2 mm
mean coil diameter	30 mm
total number of coils	15.25
outer spring	
wire diameter	8.5 mm
mean coil diameter	63 mm
total number of coils	13.25

What are most nearly the spring indices for the inner and outer springs, respectively?

(A) $C_{inner} = 3$; $C_{outer} = 5$
(B) $C_{inner} = 5$; $C_{outer} = 7$
(C) $C_{inner} = 7$; $C_{outer} = 5$
(D) $C_{inner} = 9$; $C_{outer} = 8$

SOLUTIONS

1. If the arm were locked and the ring were free to rotate, the sun and ring gears would rotate in different directions. Therefore, the velocity ratio is negative.

$$m_v = \frac{N_{ring}}{N_{sun}} = -\frac{104 \text{ teeth}}{24 \text{ teeth}} = -4.333$$

The rotational velocity of the sun is

$$\omega_{sun} = m_v \omega_{ring} + (1 - m_v)\omega_{carrier}$$

Since the ring gear is fixed, $\omega_{ring} = 0$, and the rotational velocity of the carrier is

$$\omega_{carrier} = \frac{\omega_{sun}}{1 - m_v} = \frac{50 \frac{\text{rev}}{\text{min}}}{1 - (-4.333)}$$
$$= 9.38 \text{ rpm} \quad (9.4 \text{ rpm})$$

The answer is (A).

2. Using the relationship $D = Cd$, the spring rate constant equation is

$$k = \frac{d^4 G}{8 D^3 N} = \frac{d^4 G}{8(Cd)^3 N} = \frac{dG}{8C^3 N}$$

The load deflection equation relates the properties of the spring to the applied force, where $F = pA$.

$$k = \frac{F}{x} = \frac{dG}{8C^3 N}$$

$$N = \frac{Gdx}{8C^3 F} = \frac{Gdx}{8C^3 pA}$$

$$= \frac{(83 \times 10^9 \text{ Pa})(2.5 \text{ mm})(11 \text{ mm})}{(8)(8)^3 (1.4 \times 10^6 \text{ Pa})(80 \text{ mm}^2)}$$

$$= 4.975$$

The total number of spring coils includes both the active coils and the end coils. For squared and ground ends, the total number of coils is

$$N_t = N + 2 = 4.975 + 2$$
$$= 6.975 \quad (7 \text{ coils})$$

The answer is (D).

3. Since the ring gear rotates in a different direction from gear A, the rotational velocity of the ring gear is

$$\omega_{ring} = \omega_A \left(-\frac{N_A}{N_{ring}}\right) = \left(-100 \frac{\text{rev}}{\text{min}}\right)\left(-\frac{30 \text{ teeth}}{100 \text{ teeth}}\right)$$
$$= 30 \text{ rev/min} \quad [\text{clockwise}]$$

Since the ring and sun gears rotate in different directions, the velocity ratio is negative.

$$m_v = -\frac{N_{ring}}{N_{sun}} = -\frac{80 \text{ teeth}}{40 \text{ teeth}} = -2$$

The rotational velocity of the sun gear is

$$\omega_{sun} = m_v \omega_{ring} + (1 - m_v)\omega_{carrier}$$
$$= (-2)\left(30 \frac{\text{rev}}{\text{min}}\right) + (1 - (-2))\left(60 \frac{\text{rev}}{\text{min}}\right)$$
$$= 120 \text{ rpm} \quad [\text{clockwise}]$$

The answer is (D).

4. The tangential force is

$$W_t = \frac{2T}{d} = \frac{(2)(1100 \text{ N·mm})}{19 \text{ mm}}$$
$$= 115.8 \text{ N}$$

The radial force is

$$W_r = W_t \tan \phi = (115.8 \text{ N})\tan 14.5°$$
$$= 29.9 \text{ N} \quad (30 \text{ N})$$

The answer is (B).

5. The total number of coils, N_t, for squared and ground ends is equal to $N + 2$. Therefore, for the inner spring, the number of active coils, N, is

$$N_t = N + 2$$
$$N_{inner} = N_t - 2 = 12.75 - 2 = 10.75 \quad (11)$$

For the outer spring,

$$N_{outer} = N_t - 2 = 10.25 - 2 = 8.25 \quad (9)$$

The answer is (B).

6. Using the relationship $D = Cd$ the spring rate constant equation is

$$k = \frac{d^4 G}{8 D^3 N} = \frac{d^4 G}{8(Cd)^3 N} = \frac{dG}{8C^3 N}$$

Rearranging and substituting F/x for k, the minimum wire diameter is

$$d = \frac{8FC^3 N}{Gx} = \frac{(8)(220 \text{ N})(9)^3 (12)\left(1000 \frac{\text{mm}}{\text{m}}\right)}{(83 \times 10^9 \text{ Pa})(0.012 \text{ m})}$$

$$= 15.46 \text{ mm} \quad (15 \text{ mm})$$

The answer is (A).

7. If the arm were locked and the ring were free to rotate, the sun and ring gears would rotate in different directions. Therefore, the velocity ratio is negative.

$$m_v = -\frac{64 \text{ teeth}}{32 \text{ teeth}} = -2$$

Since the ring gear is fixed, $\omega_{\text{ring}} = 0$. For this problem, assume that clockwise rotation is positive.

$$\omega_{\text{sun}} = m_v \omega_{\text{ring}} + (1 - m_v)\omega_{\text{carrier}}$$
$$100 \frac{\text{rev}}{\text{min}} = (-2)(0) + (1 - (-2))\omega_{\text{carrier}}$$
$$\omega_{\text{carrier}} = 33.3 \text{ rev/min} \quad (33 \text{ rpm})$$

The answer is (C).

8. Rearrange the equation for rotational velocity about O_4.

$$\omega_4 = \frac{a\omega_2 \sin(\theta_2 - \theta_3)}{c \sin(\theta_4 - \theta_3)}$$
$$\omega_2 = \frac{a\omega_2 \sin(\theta_2 - \theta_3)}{c \sin(\theta_4 - \theta_3)}$$
$$\sin(\theta_4 - \theta_3) = \frac{a \sin(\theta_2 - \theta_3)}{c}$$

Substitute $\theta_4 = 180° - \beta$ into the equation, and solve for β.

$$\sin(\theta_4 - \theta_3) = \frac{a \sin(\theta_2 - \theta_3)}{c}$$
$$\sin\bigl((180° - \beta) - 30°\bigr) = \frac{(3 \text{ cm})\sin(45° - 30°)}{8 \text{ cm}}$$
$$\sin(150° - \beta) = 0.097057$$
$$\beta = 144.43° \quad (140°)$$

The answer is (C).

9. Angle $\theta_4 = 180° - \alpha = 180° - 60° = 120°$. The rotational velocity about O_4 is

$$\omega_4 = \frac{a\omega_2 \sin(\theta_2 - \theta_3)}{c \sin(\theta_4 - \theta_3)}$$
$$= \frac{(2 \text{ cm})\left(-2 \frac{\text{rad}}{\text{s}}\right)\sin(45° - 30°)}{(4.8 \text{ cm})\sin(120° - 30°)}$$
$$= -0.216 \text{ rad/s} \quad (-0.22 \text{ rad/s})$$

The answer is (B).

10. For the inner spring,

$$C_{\text{inner}} = \frac{D}{d} = \frac{30 \text{ mm}}{6.2 \text{ mm}} = 4.84 \quad (5)$$

For the outer spring,

$$C_{\text{outer}} = \frac{D}{d} = \frac{63 \text{ mm}}{8.5 \text{ mm}} = 7.41 \quad (7)$$

The answer is (B).

44 Hydraulic and Pneumatic Mechanisms

PRACTICE PROBLEMS

1. In a hydraulic system, the ratio of the change in stored volume to the change in pressure is known as the fluid

(A) inductance
(B) impedance
(C) compliance
(D) inertance

2. Which formula is NOT used to calculate pipe wall thickness in a hydraulic mechanism?

(A) Boardman formula
(B) Barlow formula
(C) Lamé formula
(D) Joukowsky formula

3. Which type of valve is represented by the fluid power circuit symbol shown?

(A) nonreturn valve
(B) pressure reducing valve
(C) solenoid-actuated servo valve
(D) shut-off valve

4. A flexible hose is pressurized to more than 1.7 MPa. What is most nearly the working pressure when taken as a percentage of the burst pressure?

(A) 5.0%
(B) 10%
(C) 25%
(D) 50%

5. As it relates to hydraulic components, what does the acronym NPTF mean?

(A) Non-Pressurized Torque Fitting
(B) New Pipe Thermal Fitting
(C) National Pipe Thread for Fuel
(D) Negative Pressure Transfer Fitting

6. Generally, what is the flare angle for flare fittings used for high-pressure applications?

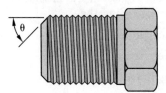

(A) 30°
(B) 37°
(C) 45°
(D) 60°

7. Which of the following hydraulic hose components generally determines the pressure at which the hose can operate?

(A) covering
(B) reinforcement
(C) end fitting
(D) tubing

8. Which of the following fitting types does NOT require O-rings?

(A) bossed
(B) flared
(C) flat-faced
(D) flanged

9. What are the basic components of a hydraulic system?

I. fluid
II. reservoir
III. pump
IV. lines
V. valves
VI. actuator

(A) I, II, III, and IV
(B) III, V, and VI
(C) III, IV, V, and VI
(D) I, II, III, IV, V, and VI

10. Which of the following components is NOT part of a pneumatic system?

(A) pump

(B) line

(C) valve

(D) actuator

11. In an existing pneumatic application, a larger force is required to overcome increased friction in an aging component. The most likely method of obtaining a larger force is increasing the

(A) working pressure

(B) cylinder size

(C) tubing/line diameter

(D) actuator stroke length

SOLUTIONS

1. The ratio of change in stored volume to the change in pressure is fluid compliance, also known as fluid capacitance.

The answer is (C).

2. The Boardman formula, the Barlow formula, and the Lamé formula are all used to calculate pipe wall thickness in a hydraulic mechanism. The Joukowsky formula describes the pressure profile of a pressure fluctuation within a piping system.

The answer is (D).

3. The symbol shown represents a nonreturn valve, also known as a check valve.

The answer is (A).

4. The working pressure is taken as 25% of the burst pressure.

The answer is (C).

5. National Pipe Thread for Fuel (NPTF) is used to designate a class of threaded fittings with tapered heads. When the male and female fittings are joined, the threads clash and form a seal that prevents fuel leakage.

The answer is (C).

6. Generally, SAE 45° flare fittings are used for low-pressure applications such as fuel lines, hot oil lines, and refrigerant lines. 45° flare fittings are commonly used in automotive and marine applications that are plumbed with copper tubing. Hydraulic equipment using higher pressures is connected by steel (not copper) tubing. A 37° seat angle was adopted because steel tubing generally cannot be flared to an angle greater than 37° without weakening it. 37° angle seats are commonly used to join hydraulic hose assemblies to hydraulic system components.

The answer is (B).

7. The strength of a hose is determined by the strength of the reinforcement embedded within the hose material.

The answer is (B).

8. Bossed, flat-faced, and flanged fittings have specific features (e.g., grooves or unthreaded lands) for holding O-rings. Flared fittings do not require O-rings.

The answer is (B).

9. The pump in a hydraulic system moves hydraulic fluid from the fluid reservoir through the lines to the actuator. Movement of the fluid is controlled by valves.

The answer is (D).

10. Hydraulic and pneumatic systems both have lines, valves, and actuators. Hydraulic include pumps to pressurize liquids. Pneumatic systems use compressors to pressurize gases.

The answer is (A).

11. The easiest way to obtain a larger actuating force in both hydraulic and pneumatic systems is to increase the actuating cylinder diameter.

The answer is (B).

45 Pressure Vessels

PRACTICE PROBLEMS

1. Which pressure vessel RT marking designation indicates a joint efficiency of 0.85?

(A) RT-1
(B) RT-2
(C) RT-3
(D) RT-4

2. Which nameplate marking indicates a pressure vessel safety valve?

(A) V
(B) NV
(C) UM
(D) UV

3. In a pressure vessel, a type 2 weld joint is a

(A) double full-fillet lap joint
(B) single-welded butt joint with a backing strip
(C) double-welded butt joint
(D) single full-fillet lap joint without plug welds

4. As it applies to pressure vessels designed in accordance with the ASME *Boiler and Pressure Vessel Code* (BPVC), what is a major difference between shakedown and ratcheting?

(A) Shakedown is elastic, and ratcheting is plastic.
(B) Shakedown is temporary, and ratcheting is permanent.
(C) Shakedown is static, and ratcheting is progressive.
(D) Shakedown is thermal, and ratcheting is mechanical.

5. The preferred fluid used in the hydrostatic testing of boilers and pressure vessels is

(A) nitrogen
(B) compressed air
(C) water
(D) hydraulic oil

6. Which marking would be applied to a power boiler designed in accordance with the BPVC?

(A) S
(B) A
(C) E
(D) M

7. What is the name of the feature used to facilitate cleaning the inside of a water boiler?

(A) inspection plate
(B) flange
(C) nozzle
(D) handhole

SOLUTIONS

1. The "RT-3" mark indicates that spot radiographic inspection was performed on all longitudinal and circumferential seams, but no nozzle connection welds were radiographed. RT-3 results in a 0.85 joint efficiency.

The answer is (C).

2. A pressure vessel safety valve is indicated by the "UV" mark.

The answer is (D).

3. A type 2 weld is a single-welded butt joint with a backing strip. Backing strips are typically removed after welding, but the backing strip in a type 2 weld is left in place.

The answer is (B).

4. Shakedown results in small permanent plastic deformations that do not increase with pressure cycling (i.e., it is static). Ratcheting produces plastic deformations that increase over time (i.e., it is progressive).

The answer is (C).

5. For safety, hydraulic testing uses liquids, not gases. To reduce costs and to avoid contaminating the pressure vessel, water is used.

The answer is (C).

6. Power boilers are marked with "S." Boiler assemblies are marked with "A." Electric boilers are marked with "E." Miniature boilers are marked with "M." Pressure piping is marked with "PP." Pressure valves are marked with "V." Heating boilers are marked with "H."

The answer is (A).

7. Handholes are openings through which a hand or equipment may be inserted for cleaning.

The answer is (D).

46 Manufacturability, Quality, and Reliability

PRACTICE PROBLEMS

1. A shaft with an interference fit has a maximum diameter of 3 cm and a nominal diameter of 2.990 cm. The upper and lower deviations of the shaft are 0.005 cm and 0.003 cm, respectively. What is most nearly the minimum shaft diameter?

(A) 2.990 cm

(B) 2.992 cm

(C) 2.993 cm

(D) 2.995 cm

2. A shaft with a clearance fit has a nominal diameter of 12 mm. The lower deviation and upper deviation are 0.036 mm and 0.028 mm, respectively. What is most nearly the maximum nominal size of the shaft?

(A) 12.028 mm

(B) 12.032 mm

(C) 12.036 mm

(D) 12.064 mm

3. A shaft has a nominal diameter of 15 mm. The shaft diameter is specified with a tolerance range of 14.950 mm to 15.027 mm. What is most nearly the tolerance of the shaft?

(A) 0.015 mm

(B) 0.023 mm

(C) 0.050 mm

(D) 0.073 mm

4. A hollow aluminum cylinder is pressed over a hollow brass cylinder as shown. Both cylinders are 5 cm long. The interference is 0.010 cm. The average coefficient of friction during assembly is 0.25. The pressure on the cylinders is 37 MPa.

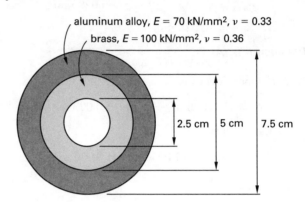

What is most nearly the initial axial disassembly force required to separate the two cylinders?

(A) 57 kN

(B) 61 kN

(C) 65 kN

(D) 73 kN

SOLUTIONS

1. The fundamental deviation, δ_F, is the smaller of the upper and lower deviations, which is 0.003 cm.

$$d_{min} = d + \delta_F = 2.990 \text{ cm} + 0.003 \text{ cm}$$
$$= 2.993 \text{ cm}$$

The answer is (C).

2. For a clearance fit, the fundamental deviation, δ_F, is the upper deviation, δ_u, which is 0.028 mm.

The maximum nominal size of the shaft is

$$d_{max} = d + \delta_F = 12 \text{ mm} + 0.028 \text{ mm}$$
$$= 12.028 \text{ mm}$$

The answer is (A).

3. The upper deviation is

$$\delta_u = 15.027 \text{ mm} - 15 \text{ mm} = 0.027 \text{ mm}$$

The lower deviation is

$$\delta_l = 15 \text{ mm} - 14.950 \text{ mm} = 0.050 \text{ mm}$$

The shaft tolerance is

$$\Delta_d = |\delta_u - \delta_l|$$
$$= |0.027 \text{ mm} - 0.050 \text{ mm}|$$
$$= 0.023 \text{ mm}$$

The answer is (B).

4. The initial force necessary to disassemble the two cylinders is the same as the maximum assembly force.

$$F_{max} = 2\pi r_{shaft} \mu p l_{interface}$$

$$= \frac{2\pi (2.5 \text{ cm})(0.25)(37 \text{ MPa})(5 \text{ cm})\left(1000 \frac{\text{Pa}}{\text{MPa}}\right)}{\left(100 \frac{\text{cm}}{\text{m}}\right)^2}$$

$$= 72.65 \text{ kN} \quad (73 \text{ kN})$$

The answer is (D).

47 Measurement and Instrumentation

PRACTICE PROBLEMS

1. It is desired to choose a resistance temperature detector (RTD) transducer for a measurement. The actual value of the temperature is not important because the initial temperature is known by other means, but the change in temperature as the test item is heated is important. Which statement is true about the selection of the RTD?

(A) The precision is important, and the accuracy is not important.

(B) The accuracy is important, and the precision is not important.

(C) Neither the accuracy nor the precision is important.

(D) Both accuracy and precision are important.

2. Using a gage factor of 2.1, what is most nearly the output sensitivity for the strain gauge bridge circuit shown?

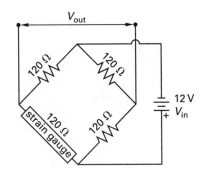

(A) 0.5 V·cm/cm
(B) 3.0 V·cm/cm
(C) 6.0 V·cm/cm
(D) 6.3 V·cm/cm

3. A resistance temperature detector (RTD) transducer is to be chosen such that the analog-to-digital conversion will have a resolution of 0.001°C. The conversion uses 16 bits. The RTD will be in a circuit with a lower voltage of -0.150 V at the lowest temperature extreme and an upper voltage of 0.300 V at the highest temperature extreme. The RTD will have an output of 0.000 V at 0.000°C. What is most nearly the sensitivity of the RTD, and what is most nearly the temperature range that the analog-to-digital conversion can represent?

(A) 96.0°C/V; -43.9°C to 98.6°C
(B) 146°C/V; -21.9°C to 43.7°C
(C) 160°C/V; -11.7°C to 38.8°C
(D) 222°C/V; 1.90°C to 78.7°C

4. The output sensitivity of a strain gauge bridge circuit is 10 V·cm/cm. If the output voltage is 10 mV, what is most nearly the strain?

(A) 0.0001 cm/cm
(B) 0.0010 cm/cm
(C) 0.0100 cm/cm
(D) 0.1000 cm/cm

5. The dimensionless incinerability index, I, of a material in a waste stream is given by

$$I = C + \frac{100\,\frac{\text{kcal}}{\text{g}}}{H}$$

In a particular waste stream, the mass percent of chlordane, C, is 0.02, and the heating value, H, is 2.71 kcal/g. The uncertainty in H is ± 0.06 kcal/g, and the uncertainty in C is $\pm 10\%$. The uncertainty in I is most nearly

(A) 0.06
(B) 0.2
(C) 0.3
(D) 0.8

SOLUTIONS

1. Precision is a measure of the repeatability of results, which is important for this measurement. The accuracy can have a significant bias, and the RTD would be acceptable for this measurement, so the accuracy is not important.

The answer is (A).

2. Find the change in resistance, ΔR.

$$GF = \frac{\frac{\Delta R}{R}}{\frac{\Delta L}{L}} = \frac{\frac{\Delta R}{R}}{\varepsilon}$$

$$\Delta R = \varepsilon(R)(GF)$$

Substitute the equation for ΔR into the equation for reference voltage.

$$V_0 \approx \left(\frac{\Delta R}{4R}\right) \cdot V_{in}$$

$$= \left(\frac{\varepsilon(R)(GF)}{4R}\right) V_{in}$$

$$\frac{V_0}{\varepsilon} = \frac{V_{IN}(GF)}{4} = \frac{(12 \text{ V})(2.1)}{4}$$

$$= 6.3 \text{ V·cm/cm}$$

(Any linear unit can be used for strain.)

The answer is (D).

3. The resolution of the circuit is

$$\varepsilon_V = \frac{V_H - V_L}{2^n} = \frac{0.300 \text{ V} - (-0.150 \text{ V})}{2^{16}}$$

$$= 6.8665 \times 10^{-6} \text{ V}$$

This voltage is to represent $0.001°C$ with the RTD; therefore, the sensitivity of the RTD should be

$$\frac{0.001°C}{6.8665 \times 10^{-6} \text{ V}} = 145.64°C/V \quad (146°C/V)$$

The temperature range that this sensor can represent is derived from the number of bits and the value that represents $0°C$. The voltage range is

$$(0.001°C)(2^{16}) = 65.54°C$$

The value that represents $0°C$ is 0 V, and one third of the voltage range is below zero while two thirds of the voltage range is above zero. The lower and upper limits the sensor can represent are

$$\frac{-65.54°C}{3} = -21.85°C \quad (-21.9°C)$$

$$\frac{(2)(65.54°C)}{3} = 43.69°C \quad (43.7°C)$$

The answer is (B).

4. The strain is

$$\text{sensitivity} = \frac{V_0}{\varepsilon}$$

$$\varepsilon = \frac{V_0}{\text{sensitivity}}$$

$$= \frac{10 \text{ mV}}{\left(10 \frac{\text{V·cm}}{\text{cm}}\right)\left(1000 \frac{\text{mV}}{\text{V}}\right)}$$

$$= 0.0010 \text{ cm/cm}$$

The answer is (B).

5. Use the Kline-McClintock equation.

$$w_R = \sqrt{\left(w_1 \frac{\partial f}{\partial x_1}\right)^2 + \left(w_2 \frac{\partial f}{\partial x_2}\right)^2 + \cdots + \left(w_n \frac{\partial f}{\partial x_n}\right)^2}$$

For the incinerability index,

$$I = C + \frac{100 \frac{\text{kcal}}{\text{g}}}{H}$$

$$x_1 = C$$

$$x_2 = H$$

$$w_1 = w_C = (0.1)(0.02) = 0.002$$

$$w_2 = w_H = 0.06 \text{ kcal/g}$$

$$w_R = w_I$$

$$\frac{\partial f}{\partial C} = \frac{\partial C}{\partial C} = 1$$

$$\frac{\partial f}{\partial H} = \frac{\partial \left(\frac{100 \frac{\text{kcal}}{\text{g}}}{H}\right)}{\partial H} = -\frac{100 \frac{\text{kcal}}{\text{g}}}{H^2}$$

$$w_I = \sqrt{\left((0.002)(1)\right)^2 + \left(\left(0.06 \frac{\text{kcal}}{\text{g}}\right)\left(-\frac{100 \frac{\text{kcal}}{\text{g}}}{\left(2.71 \frac{\text{kcal}}{\text{g}}\right)^2}\right)\right)^2}$$

$$= 0.817 \quad (0.8)$$

This means that I is 38.9 ± 0.8.

The answer is (D).

48 Controls

PRACTICE PROBLEMS

1. The frequency response of a system to a sinusoidal input is given by

$$\frac{M}{\alpha} = \frac{X_2(j\omega)}{X_1(j\omega)}$$

By differentiation, the peak value of M, M_p, and the frequency at which it occurs, ω_r, are expressed in terms of the damping ratio, ζ, and natural frequency, ω_n.

$$M_{\max} = \frac{1}{2\zeta\sqrt{1-\zeta^2}}$$

$$\omega_r = \omega_n\sqrt{1-2\zeta^2}$$

Which curve in the illustration shown best represents the frequency response?

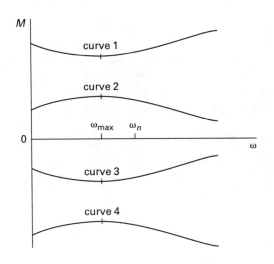

(A) curve 1
(B) curve 2
(C) curve 3
(D) curve 4

2. For the given block diagram, K is set to 16.

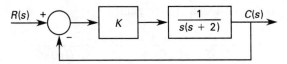

The damped natural frequency of oscillation in response to a step input, ω_d, is most nearly

(A) 3.9
(B) 10
(C) 12
(D) 16

3. The frequency response of a second-order system is given by

$$\frac{M}{\alpha} = \frac{X_2(j\omega)}{X_1(j\omega)}$$

The peak value of M, M_p, and the frequency at which it occurs, ω_r, are expressed in terms of the damping ratio, ζ, and natural frequency, ω_n.

$$M_p = \frac{1}{2\zeta\sqrt{1-\zeta^2}}$$

$$\omega_r = \omega_n\sqrt{1-2\zeta^2}$$

A polar plot of the system is shown.

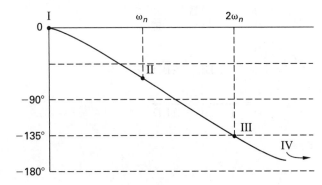

Where does the peak response in amplitude occur?

(A) point II
(B) point IV
(C) between points I and II
(D) between points II and III

4. A block diagram is shown.

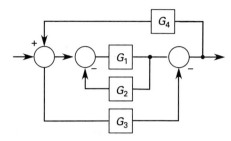

What is the overall system gain?

(A) $\dfrac{G_1 + G_2 + G_1 G_2}{1 + G_1 G_3 + G_1 G_2 + G_2 G_3 + G_1 G_2 G_3 G_4}$

(B) $\dfrac{G_1 + G_4 + G_2 G_3 G_4}{1 + G_1 G_3 + G_1 G_2 + G_2 G_3 + G_1 G_2 G_3 G_4}$

(C) $\dfrac{G_1 - G_2 + G_3 G_4}{1 + G_1 G_2 + G_1 G_4 + G_3 G_4 + G_1 G_2 G_3 G_4}$

(D) $\dfrac{G_1 + G_3 + G_1 G_2 G_3}{1 + G_1 G_2 - G_1 G_4 - G_3 G_4 - G_1 G_2 G_3 G_4}$

5. Simplify the following block diagram.

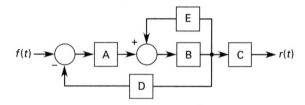

What is the overall system gain?

(A) $\dfrac{r(t)}{f(t)} = \dfrac{ABE}{1 + BC - ABE}$

(B) $\dfrac{r(t)}{f(t)} = \dfrac{ABE}{1 - BC + ABE}$

(C) $\dfrac{r(t)}{f(t)} = \dfrac{ABC}{1 + BE - ABD}$

(D) $\dfrac{r(t)}{f(t)} = \dfrac{ABC}{1 - BE + ABD}$

6. Simplify the following block diagram.

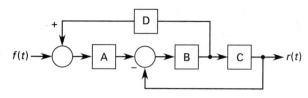

What is the overall system gain?

(A) $\dfrac{r(t)}{f(t)} = \dfrac{ABD}{1 + ABC - BD}$

(B) $\dfrac{r(t)}{f(t)} = \dfrac{ABD}{1 - ABC + BD}$

(C) $\dfrac{r(t)}{f(t)} = \dfrac{ABC}{1 + ABD - BC}$

(D) $\dfrac{r(t)}{f(t)} = \dfrac{ABC}{1 - ABD + BC}$

7. A block diagram is shown.

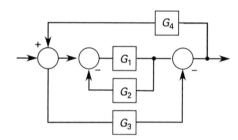

What is the system sensitivity if $G_1 = -5$ dB, $G_2 = 2$ dB, $G_3 = 4$ dB, and $G_4 = 3$ dB?

(A) $-3/34$ dB^{-1}

(B) $-3/28$ dB^{-1}

(C) $3/44$ dB^{-1}

(D) $9/31$ dB^{-1}

8. A control system with negative feedback is shown. What is $E(s)$?

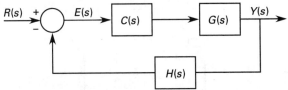

(A) $E(s) = R(s) - Y(s)$

(B) $E(s) = C(s)G(s)R(s)$

(C) $E(s) = \dfrac{G(s)R(s)}{1 + C(s)H(s)}$

(D) $E(s) = \dfrac{R(s)}{1 + \bigl(C(s)G(s)\bigr)H(s)}$

9. A control system with negative feedback is shown.

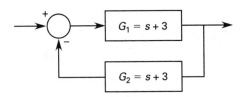

What is the time-domain transfer function of the system?

(A) $e^{-3t}\sin t$

(B) $e^{-3t}\cos t$

(C) $\dfrac{t+3}{1+(t+3)^2}$

(D) $\dfrac{1}{1+(t+3)^2}$

10. The transfer function for a control system is

$$T(s) = \dfrac{b_0 s^p + b_1 s^{p-1} + \cdots + b_p}{s^n + a_1 s^{n-1} + \cdots + a_n} \quad [a_n \neq 0]$$

What is the steady-state response to a step input for the control system?

(A) b_p/a_n

(B) b_p^2/a_n

(C) b_p/a_n^2

(D) b_p

11. A control system with negative feedback is constructed from linear time-invariant elements as shown.

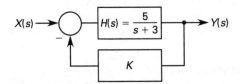

What is the requirement of the constant K so that the closed-loop system is stable?

(A) $K \leq -5/3$

(B) $K \leq -3/5$

(C) $K \geq -3/5$

(D) $K \geq 0$

SOLUTIONS

1. As ω increases to ω_r, M should increase to its peak value. The shape of curve 2 and curve 4 are both correct. However, since M_p is always positive, only curve 2 is correct.

The answer is (B).

2. The general characteristic equation is

$$s^2 + 2\zeta\omega_n s + \omega_n^2 = 0$$

The characteristic equation for this system is

$$s^2 + 2s + 16 = 0$$

Find ω_n and ζ.

$$\omega_n = \sqrt{16} = 4$$

$$\zeta = \frac{2s}{2\omega_n s} = \frac{1}{\omega_n} = \frac{1}{4}$$
$$= 0.25$$

The damped natural frequency of oscillation is

$$\omega_d = \omega_n\sqrt{1-\zeta^2}$$
$$= 4\sqrt{1-(0.25)^2}$$
$$= 3.87 \quad (3.9)$$

The answer is (A).

3. As stated in the problem statement, the peak occurs at ω_r. From the equation for ω_r, when $\zeta = 0$, the peak occurs at $\omega_r = \omega_n$. When $\zeta > 0$, the peak occurs at $\omega_r < \omega_n$ (i.e., between points I and II).

The answer is (C).

4. Simplify the block diagram.

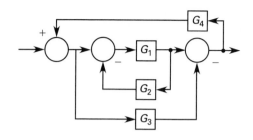

step 1

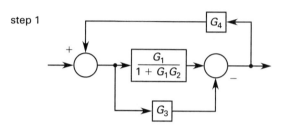

step 2

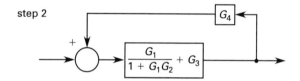

step 3

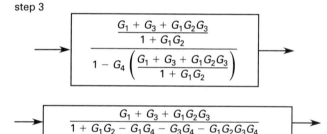

The overall system gain is

$$\frac{G_1 + G_3 + G_1 G_2 G_3}{1 + G_1 G_2 - G_1 G_4 - G_3 G_4 - G_1 G_2 G_3 G_4}$$

The answer is (D).

5. Simplify the block diagram.

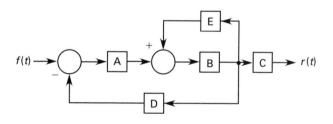

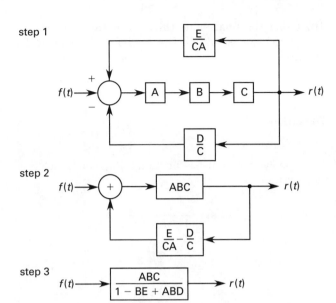

The overall system gain is

$$\frac{r(t)}{f(t)} = \frac{ABC}{1 - BE + ABD}$$

The answer is (D).

6. Simplify the block diagram.

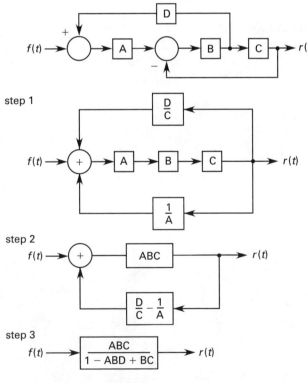

The overall system gain is

$$\frac{r(t)}{f(t)} = \frac{ABC}{1 - ABD + BC}$$

The answer is (D).

7. Simplify the block diagram.

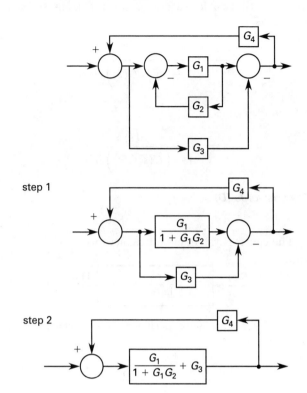

From step 2 of the simplified block diagram,

$$G(s) = \frac{G_1}{1 + G_1 G_2} + G_3$$

$$= \frac{-5 \text{ dB}}{1 + (-5 \text{ dB})(2 \text{ dB})} + 4 \text{ dB}$$

$$= 41/9 \text{ dB}$$

$$H(s) = G_4$$

$$= 3 \text{ dB}$$

$$G(s)H(s) = \left(\frac{41}{9} \text{ dB}\right)(3 \text{ dB})$$

$$= 41/3 \text{ dB} > 0$$

The system has negative feedback. The sensitivity of the system is

$$S = \frac{1}{1 + G(s)H(s)}$$

$$= \frac{1}{1 + \frac{41}{3} \text{ dB}}$$

$$= 3/44 \text{ dB}^{-1}$$

The answer is (C).

8. The error ratio for a closed-loop, negative feedback system with two forward-loop controllers in series is defined as

$$\frac{E(s)}{R(s)} = \frac{1}{1 + \big(C(s)G(s)\big)H(s)}$$

Therefore,

$$E(s) = \frac{R(s)}{1 + \big(C(s)G(s)\big)H(s)}$$

The answer is (D).

9. The control system circuit reduces to

$$R(s) \rightarrow \boxed{\frac{G_1}{1 \pm G_1 G_2}} \rightarrow Y(s)$$

The problem statement specifies this is a negative feedback loop, so

$$\frac{Y(s)}{R(s)} = \frac{s+3}{1+(s+3)^2}$$

The inverse Laplace transform is

$$\frac{C(t)}{R(t)} = \mathcal{L}^{-1}\left(\frac{s+3}{1+(s+3)^2}\right) = e^{-3t}\cos t$$

The answer is (B).

10. Using the final value theorem, obtain the steady-state step response by substituting 0 for s in the transfer function.

$$Y(s) = \lim_{s \to 0} T(s) = T(0) = b_p/a_n$$

The answer is (A).

11. The system transfer function, $G(s)$, for the closed-loop system is

$$G(s) = \frac{H(s)}{1 + KH(s)}$$

$$= \frac{\frac{5}{s+3}}{1 + K\left(\frac{5}{s+3}\right)}$$

$$= \frac{5}{s+3+5K}$$

In order for the system to be stable, the pole of the closed-loop system transfer function needs to be at the left-hand side of the plane (i.e., must be negative).

$$s_1 = -3 - 5K \leq 0$$

$$K \geq -3/5$$

The answer is (C).

49 Computer Software

PRACTICE PROBLEMS

1. Which of the following flowcharts does NOT represent a complete program?

(A)

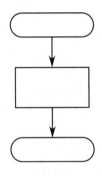

(B)

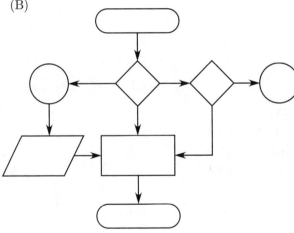

(C)

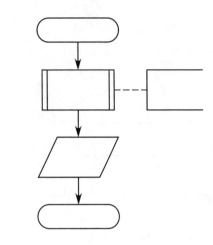

(D)

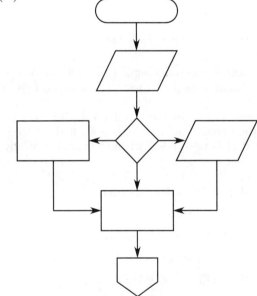

2. What flowchart element would be used to represent an IF...THEN statement?

(A)

(B)

(C)

(D)

3. In programming, a recursive function is one that

(A) calls previously used functions
(B) generates functional code to replace symbolic code
(C) calls itself
(D) compiles itself in real time

4. Structured programming is to be used to determine whether examinees pass a test. A passing score is 70 or more out of a possible 100. Which of the following IF statements would set the variable PASSED to 1 (true) when the variable SCORE is passing, and set the variable PASSED to 0 (false) when the variable SCORE is not passing?

(A) IF SCORE > 70 PASSED = 1
    ELSE PASSED = 0
(B) IF SCORE > 69 PASSED = 1
    ELSE PASSED = 0
(C) IF SCORE < 69 PASSED = 1
(D) IF SCORE < 69 PASSED = 0
    ELSE PASSED = 1

5. A structured programming segment contains the following program segment. What is the value of Y after the segment is executed?

$$Y = 4$$
$$B = 4$$
$$Y = 3*B - 6$$
$$\text{IF } Y > B \text{ THEN } Y = B - 2$$
$$\text{IF } Y < B \text{ THEN } Y = Y + 2$$
$$\text{IF } Y = B \text{ THEN } Y = B + 2$$

(A) 2
(B) 6
(C) 8
(D) 12

6. The structured programming segment shown implements which of the following equations? The variable N is an integer greater than 0.

$$A = X$$
$$\text{DO UNTIL } N = 0$$
$$Y = A*X$$
$$A = Y$$
$$N = N - 1$$
$$\text{END UNTIL}$$

(A) $Y = X!$
(B) $Y = X^{N+1}$
(C) $Y = X^{N-1}$
(D) $Y = X^N$

7. In the structured programming fragment shown, what can line 2 be accurately described as?

```
1   REAL X, Y
2   X = 3
3   Y = COS(X)
4   PRINT Y
```

(A) an assignment
(B) a command
(C) a declaration
(D) a function

8. In a typical spreadsheet program, what cell is directly below cell AB4?

(A) AB5

(B) AC4

(C) AC5

(D) BC4

9. Which of the following terms is best defined as a formula or set of steps for solving a particular problem?

(A) program

(B) software

(C) firmware

(D) algorithm

10. Which of the following is the computer language that is executed within a computer's central processing unit?

(A) MS-DOS

(B) high-level language

(C) assembly language

(D) machine language

11. Which of the following best defines a compiler?

(A) hardware that is used to translate high-level language to machine code

(B) software that collects and stores executable commands in a program

(C) software that is used to translate high-level language into machine code

(D) hardware that collects and stores executable commands in a program

12. The effect of using recursive functions in a program is generally to

(A) use less code and less memory

(B) use less code and more memory

(C) use more code and less memory

(D) use more code and more memory

13. When can an 8-bit system correctly access more than 128 different integers?

(A) when the integers are in the range of $[-255, 0]$

(B) when the integers are in the range of $[0, 256]$

(C) when the integers are in the range of $[-128, 128]$

(D) when the integers are in the range of $[0, 512]$

14. What computer operating system (OS) is required to view a document saved in HTML (hypertext markup language) format?

I. Apple OS

II. MS-DOS

III. Windows

IV. Unix

(A) I or III

(B) I, II, or III

(C) I, III, or IV

(D) I, II, III, or IV

15. A typical spreadsheet for economic evaluation of alternatives uses cell F4 to store the percentage value of inflation rate. The percentage rate is assumed to be constant throughout the lifetime of the study. What variable should be used to access that value throughout the model?

(A) F4

(B) $F4

(C) %F4

(D) F4

16. Refer to the following portion of a spreadsheet.

	A	B	C
1	10	11	12
2	1	A2^2	
3	2	A3^2	
4	3	A4^2	
5	4	A5^2	

The top-to-bottom values in column B will be

(A) 11, 1, 2, 3, 4

(B) 11, 1, 3, 6, 10

(C) 11, 1, 4, 9, 16

(D) 11, 1, 5, 12, 22

17. Refer to the following portion of a spreadsheet.

	A	B	C	D
1	10	11	12	13
2	5	4	B2*A$1	
3	6	5	B3*B$1	
4	7	6	B4*C$1	
5	8	7	B5*D$1	

The top-to-bottom values in column C will be

- (A) 12, 20, 30, 42, 56
- (B) 12, 40, 55, 72, 91
- (C) 12, 50, 66, 84, 104
- (D) 12, 100, 121, 144, 169

SOLUTIONS

1. A flowchart must begin and end with a terminal symbol. The symbol at the bottom of answer D is the "off-page" symbol, which indicates that the flowchart continues on the next page. This is not a complete program.

The answer is (D).

2. An IF...THEN statement alters the flow of a program based on some criterion that can be evaluated as true or false. The diamond-shaped symbol is used to represent a decision.

The answer is (B).

3. A recursive function calls itself.

The answer is (C).

4. Answer A will not give the correct response when SCORE = 70. Answer C will not set PASSED to 0. Answer D will not give the correct response when SCORE = 69. Answer B will set PASSED to 1 for SCORE = 70 to 100 and will set PASSED to 0 for SCORE = 0 to 69.

The answer is (B).

5. The first operation changes the value of Y.

$$Y = 3 \times 4 - 6 = 6$$

The first IF statement is satisfied, so the operation is performed.

$$Y = 4 - 2 = 2$$

However, the program execution does not end here.

The value of Y is then less than B, so the second IF statement is executed. This statement is satisfied, so the operation is performed.

$$Y = 2 + 2 = 4$$

The value of Y is then equal to B, so the third IF statement is executed. This statement is satisfied, so the operation is performed.

$$Y = 4 + 2 = 6$$

The answer is (B).

6. The DO loop will be executed N times. After the first execution, $Y = X^2$. After each subsequent execution of the loop, Y is multiplied by X. Therefore, the segment calculates $Y = X^{N+1}$.

The answer is (B).

7. Line 2 is an assignment. A command directs the computer to take some action, such as PRINT. A declaration states what type of data a variable will contain (like REAL) and reserves space for it in memory. A function performs some specific operation (like finding the COSine of a number) and returns a value to the program.

The answer is (A).

8. Spreadsheets generally label a cell by giving its column and row, in that order. Cell AB4 is in column AB, row 4. The cell directly below AB4 is in column AB, row 5, designated as AB5.

The answer is (A).

9. A program is a sequence of instructions that implements a formula or set of steps but is not a formula or set of steps. Software and firmware are programs on media. An algorithm is a formula or set of steps for solving a particular problem that is often implemented as a program.

The answer is (D).

10. The central processing unit executes a version of the program that has been compiled into the machine language and that is in the form of operations and operands specific to the machine's coding.

The answer is (D).

11. A compiler is a program (i.e., software) that converts programs written in higher-level languages to lower-level languages that the computer can understand.

The answer is (C).

12. A recursive function calls itself. Since the function does not need to be coded in multiple places, less code is used. Each subsequent call of the function must be carried out in a different location, so more memory is used.

The answer is (B).

13. An 8-bit system can represent $(2)^8 = 256$ different distinct integers. Normally, the 8th bit is used for the sign, and only 7 bits are used for magnitude, resulting in a range of $[-127, 128]$ or $[-128, 127]$ (counting zero as one of the integers). If all of the integers are known or assumed to have the same sign, a range of 256 integers is available. All of the answer choices except A contain more than 256 distinct integers.

The answer is (A).

14. HTML may be viewed on any computer with a compatible browser.

The answer is (D).

15. The dollar sign symbol, "$", is used in spreadsheets to "fix" the column and/or row designator following it when other columns or rows are permitted to vary.

The answer is (D).

16. Except for the first entry (which is 11), column B calculates the square of the values in column A. The entries are 11, $(1)^2$, $(2)^2$, $(3)^2$, and $(4)^2$.

The answer is (C).

17. Except for the first entry (which is 12), column C is found by taking the numbers from column B and then multiplying by the entries in row 1. For example, B2*A$1 means to multiply the entry in B2, given as 4 in the problem statement, by the number entered in cell A1, which is 10. This product is 40.

The entries are 12, 4×10, 5×11, 6×12, 7×13, or 12, 40, 55, 72, 91.

The answer is (B).

50 Engineering Economics

PRACTICE PROBLEMS

1. Permanent mineral rights on a parcel of land are purchased for an initial lump-sum payment of $100,000. Profits from mining activities are $12,000 each year, and these profits are expected to continue indefinitely. The interest rate earned on the initial investment is most nearly

(A) 8.3%

(B) 9.0%

(C) 10%

(D) 12%

2. $1000 is deposited in a savings account that pays 6% annual interest, and no money is withdrawn for three years. The account balance after three years is most nearly

(A) $1120

(B) $1190

(C) $1210

(D) $1280

3. An oil company is planning to install a new 80 mm pipeline to connect storage tanks to a processing plant 1500 m away. The connection will be needed for the foreseeable future. An annual interest rate of 8% is assumed, and annual maintenance and pumping costs are considered to be paid in their entireties at the end of the years in which their costs are incurred.

initial cost	$1500
service life	12 yr
salvage value	$200
annual maintenance	$400
pump cost/hour	$2.50
pump operation	600 hr/yr

The capitalized cost of running and maintaining the 80 mm pipeline is most nearly

(A) $15,000

(B) $20,000

(C) $24,000

(D) $27,000

4. New 200 mm diameter pipeline is installed over a distance of 1000 m. Annual maintenance and pumping costs are considered to be paid in their entireties at the end of the years in which their costs are incurred. The pipe has the following costs and properties.

initial cost	$1350
annual interest rate	6%
service life	6 yr
salvage value	$120
annual maintenance	$500
pump cost/hour	$2.75
pump operation	2000 hr/yr

What is most nearly the equivalent uniform annual cost (EUAC) of the pipe?

(A) $5700

(B) $5900

(C) $6100

(D) $6300

5. New 120 mm diameter pipeline is installed over a distance of 5000 m. Annual maintenance and pumping costs are considered to be paid in their entireties at the end of the years in which their costs are incurred. The pipe has the following costs and properties.

initial cost	$2500
annual interest rate	10%
service life	12 yr
salvage value	$300
annual maintenance	$300
pump cost/hour	$1.40
pump operation	600 hr/yr

What is most nearly the equivalent uniform annual cost (EUAC) of the pipe?

(A) $1200

(B) $1300

(C) $1400

(D) $1500

6. A construction company purchases 100 m of 40 mm diameter steel cable with an initial cost of $4500. The annual interest rate is 4%, and annual maintenance costs are considered to be paid in their entireties at the end of the years in which their costs are incurred. The annual maintenance cost of the cable is $200/yr over a service life of nine years. Using Modified Accelerated Cost Recovery System (MACRS) depreciation and assuming a seven year recovery period, the depreciation allowance for the cable in the first year of operation is most nearly

(A) $640

(B) $670

(C) $720

(D) $860

7. A piece of equipment has an initial cost of $5000 in year 1. The maintenance cost is $300/yr for the total lifetime of seven years. During years 1–3, the rate of inflation is 5%, and the effective annual rate of interest is 9%. The uninflated present worth of the equipment during year 1 is most nearly

(A) $3200

(B) $3300

(C) $3400

(D) $3500

8. A company is considering buying a computer with the following costs and interest rate.

initial cost	$3900
salvage value	$1800
useful life	10 years
annual maintenance	$390
interest rate	6%

The equivalent uniform annual cost (EUAC) of the computer is most nearly

(A) $740

(B) $780

(C) $820

(D) $850

9. A computer with a useful life of 13 years has the following costs and interest rate.

initial cost	$5500
salvage value	$3100
annual maintenance	
years 1–8	$275
years 9–13	$425
interest rate	6%

The equivalent uniform annual cost (EUAC) of the computer is most nearly

(A) $730

(B) $780

(C) $820

(D) $870

10. A computer with an initial cost of $1500 and an annual maintenance cost of $500/yr is purchased and kept indefinitely without any change in its annual maintenance costs. The interest rate is 4%. The present worth of all expenditures is most nearly

(A) $12,000

(B) $13,000

(C) $14,000

(D) $15,000

11. A computer with a useful life of 12 years has an initial cost of $2300 and a salvage value of $350. The interest rate is 6%. Using the straight line method, the total depreciation of the computer for the first five years is most nearly

(A) $760

(B) $810

(C) $830

(D) $920

12. A computer with a useful life of 12 years has an initial cost of $3200 and a salvage value of $100. The interest rate is 10%. Using the Modified Accelerated Cost Recovery System (MACRS) method of depreciation and a 10 year recovery period, what is most nearly the book value of the computer after the second year?

(A) $1900

(B) $2100

(C) $2300

(D) $2400

13. A computer with a useful life of five years has an initial cost of $6000. The salvage value is $2300, and the annual maintenance is $210/yr. The interest rate is 8%. What is most nearly the present worth of the costs for the computer?

(A) $5200
(B) $5300
(C) $5600
(D) $5700

14. A company must purchase a machine that will be used over the next eight years. The purchase price is $10,000, and the salvage value after eight years is $1000. The annual insurance cost is 2% of the purchase price, the electricity cost is $300 per year, and maintenance and replacement parts cost $100 per year. The effective annual interest rate is 6%. Neglect taxes. The effective uniform annual cost (EUAC) of ownership is most nearly

(A) $1200
(B) $2100
(C) $2200
(D) $2300

15. A company purchases a piece of equipment for $15,000. After nine years, the salvage value is $900. The annual insurance cost is 5% of the purchase price, the electricity cost is $600/yr, and the maintenance and replacement parts cost is $120/yr. The effective annual interest rate is 10%. Neglecting taxes, what is most nearly the present worth of the equipment if it is expected to save the company $4500 per year?

(A) $2300
(B) $2800
(C) $3200
(D) $3500

SOLUTIONS

1. Use the capitalized cost equation to find the interest rate earned.

$$P = \frac{A}{i}$$

$$i = \frac{A_{\text{profit}}}{P_{\text{cost}}} = \frac{\$12,000}{\$100,000}$$

$$= 0.12 \quad (12\%)$$

The answer is (D).

2. Find the future worth of $1000.

$$F = P(1+i)^n = (\$1000)(1+0.06)^3$$
$$= \$1191 \quad (\$1190)$$

The answer is (B).

3. The annual cost of running and maintaining the 80 mm pipeline is

$$A = \left(\frac{\$2.50}{\text{hr}}\right)(600 \text{ hr}) + \$400 = \$1900$$

Capitalized costs are the present worth of an infinite cash flow.

$$P = \frac{A}{i} = \frac{\$1900}{0.08}$$
$$= \$23,750 \quad (\$24,000)$$

The answer is (C).

4. The equivalent uniform annual cost (EUAC) is the uniform annual amount equivalent of all cash flows. When calculating the EUAC, costs are positive and income is negative.

$$\text{EUAC}_{80} = A_{\text{initial}} + A_{\text{maintenance}}$$
$$+ A_{\text{pump}} - A_{\text{salvage}}$$
$$= (\$1350)(A/P, 6\%, 6) + \$500$$
$$+ \left(\frac{\$2.75}{\text{hr}}\right)(2000 \text{ hr})$$
$$- (\$120)(A/F, 6\%, 6)$$
$$= (\$1350)(0.2034) + \$500$$
$$+ \$5500 - (\$120)(0.1434)$$
$$= \$6257 \quad (\$6300)$$

The answer is (D).

5. The equivalent uniform annual cost (EUAC) is the uniform annual amount equivalent of all cash flows. When calculating the EUAC, costs are positive and income is negative.

$$\text{EUAC}_{120} = A_{\text{initial}} + A_{\text{maintenance}}$$
$$+ A_{\text{pump}} - A_{\text{salvage}}$$
$$= (\$2500)(A/P, 10\%, 12) + \$300$$
$$+ \left(\frac{\$1.40}{\text{hr}}\right)(600 \text{ hr})$$
$$- (\$300)(A/F, 10\%, 12)$$
$$= (\$2500)(0.1468) + \$300 + \$840$$
$$- (\$300)(0.0468)$$
$$= \$1492.96 \quad (\$1500)$$

The answer is (D).

6. The MACRS factor for the first year, given a 7 year recovery period, is 14.29%.

$$D_1 = (\text{factor}) C$$
$$= (0.1429)(\$4500)$$
$$= \$643 \quad (\$640)$$

The answer is (A).

7. If the unadjusted interest rate is used to calculate the present worth, the answer will be in dollars affected by three years of inflation. To find the uninflated worth three years ago, the effect of inflation during those years must be eliminated from the calculation. To find the answer in uninflated dollars, determine the interest rate adjusted for inflation.

$$d = i + f + (i \times f)$$
$$= 0.09 + 0.05 + (0.09)(0.05)$$
$$= 0.1445$$

Use this adjusted rate in the single payment present worth equation, substituting d for i.

$$P = F(1 + d)^{-n}$$
$$= (\$5000)(1 + 0.1445)^{-3}$$
$$= \$3335 \quad (\$3300)$$

The answer is (B).

8. The equivalent uniform annual cost (EUAC) is the uniform annual amount equivalent to all cash flows. When calculating the EUAC, costs are positive and income is negative.

$$\text{EUAC} = A_{\text{initial}} + A_{\text{maintenance}} - A_{\text{salvage}}$$
$$= (\$3900)(A/P, 6\%, 10) + \$390$$
$$- (\$1800)(A/F, 6\%, 10)$$
$$= (\$3900)(0.1359) + \$390$$
$$- (\$1800)(0.0759)$$
$$= \$783 \quad (\$780)$$

The answer is (B).

9. The equivalent uniform annual cost (EUAC) is the uniform annual amount equivalent to all cash flows. When calculating the EUAC, costs are positive and income is negative. An expedient way to find the annual worth of the maintenance for the computer is to divide the maintenance costs into two annual series, one of $275 lasting from year 1 to year 13, and one of $150 (the difference between $425 and $275) lasting from year 9 to year 13. Find the future value in year 13 for each series, add them, and then convert the result back into a single annual amount.

$$F_{\$275} = A(F/A, 6\%, 13) = (\$275)(18.8821)$$
$$= \$5192.60$$
$$F_{\$150} = A(F/A, 6\%, 5) = (\$150)(5.6371)$$
$$= \$845.60$$
$$F_{\text{maintenance}} = F_{\$275} + F_{\$150} = \$5192.60 + \$845.60$$
$$= \$6038$$
$$A_{\text{maintenance}} = F_{\text{maintenance}}(A/F, 6\%, 13)$$
$$= (\$6038)(0.0530)$$
$$= \$320$$

Now calculate the EUAC.

$$\text{EUAC} = A_{\text{initial}} + A_{\text{maintenance}} - A_{\text{salvage}}$$
$$= (\$5500)(A/P, 6\%, 13) + \$320$$
$$- (\$3100)(A/F, 6\%, 13)$$
$$= (\$5500)(0.1130) + \$320$$
$$- (\$3100)(0.0530)$$
$$= \$777 \quad (\$780)$$

The answer is (B).

10. The expenditures for the computer are the initial cost of $1500 and the annual maintenance cost of $500. The annual costs continue indefinitely, so find the present worth of an infinite cash flow.

$$P_{\text{maintenance}} = \frac{A}{i} = \frac{\$500}{0.04} = \$12,500$$

The present worth of all expenditures is

$$P_{\text{total}} = P_{\text{initial}} + P_{\text{annual}} = \$1500 + \$12,500$$
$$= \$14,000$$

The answer is (C).

11. With the straight line method, the depreciation is the same every year. Find the annual depreciation.

$$D_j = \frac{C - S_n}{n} = \frac{\$2300 - \$350}{12} = \$162.50$$

The total depreciation for five years is

$$\sum D_{1-5} = (5)(\$162.50) = \$812.50 \quad (\$810)$$

The answer is (B).

12. Subtract the first two years' depreciation from the original cost.

year	factor (%)	D_j
1	10.00	$(0.10)(\$3200) = \$320$
2	18.00	$(0.18)(\$3200) = \underline{\$576}$
		$\sum D_j = \$896$

The book value is

$$\text{BV} = \text{initial cost} - \sum D_j = \$3200 - \$896$$
$$= \$2304 \quad (\$2300)$$

The answer is (C).

13. Bring all costs and benefits into the present.

$$P_{\text{total}} = P_{\text{initial}} + P_{\text{maintenance}} - P_{\text{salvage}}$$
$$= \$6000 + (\$210)(P/A, 8\%, 5)$$
$$\quad - (\$2300)(P/F, 8\%, 5)$$
$$= \$6000 + (\$210)(3.9927)$$
$$\quad - (\$2300)(0.6806)$$
$$= \$5273 \quad (\$5300)$$

The answer is (B).

14. The effective uniform annual cost (EUAC) is the annual cost equivalent of all costs. When calculating the EUAC, costs are positive and income is negative. Find the annual equivalents of all costs and add them together to get the EUAC.

$$\text{EUAC} = C_{\text{initial}}(A/P, 6\%, 8)$$
$$\quad + A_{\text{electricity}} + A_{\text{maintenance}}$$
$$\quad + A_{\text{insurance}} - S_8(A/F, 6\%, 8)$$
$$= (\$10,000)(0.1610) + \$300 + \$100$$
$$\quad + (0.02)(\$10,000)$$
$$\quad - (\$1000)(0.1010)$$
$$= \$2109 \quad (\$2100)$$

The answer is (B).

15. Add the present worths of all cash flows.

$$P_{\text{total}} = -C_{\text{initial}} - A_{\text{electricity}}(P/A, 10\%, 9)$$
$$\quad - A_{\text{maintenance}}(P/A, 10\%, 9)$$
$$\quad - A_{\text{insurance}}(P/A, 10\%, 9)$$
$$\quad + A_{\text{benefits}}(P/A, 10\%, 9)$$
$$\quad + S_9(P/F, 10\%, 9)$$
$$= -\$15,000 - (\$600)(5.7590)$$
$$\quad - (\$120)(5.7590)$$
$$\quad - (0.05)(\$15,000)(5.7590) + (\$4500)(5.7590)$$
$$\quad + (\$900)(0.4241)$$
$$= \$2831 \quad (\$2800)$$

The answer is (B).

51 Professional Practice

PRACTICE PROBLEMS

1. What must be proven for damages to be collected from a strict liability in tort?

(A) that willful negligence caused an injury

(B) that willful or unwillful negligence caused an injury

(C) that the manufacturer knew about a product defect before the product was released

(D) none of the above

2. A material breach of contract occurs when the

(A) contractor uses material not approved by the contract for use

(B) contractor's material order arrives late

(C) owner becomes insolvent

(D) contractor installs a feature incorrectly

3. If a contract has a value engineering clause and a contractor suggests to the owner that a feature or method be used to reduce the annual maintenance cost of the finished project, what will be the most likely outcome?

(A) The contractor will be able to share one time in the owner's expected cost savings.

(B) The contractor will be paid a fixed amount (specified by the contract) for making a suggestion, but only if the suggestion is accepted.

(C) The contract amount will be increased by some amount specified in the contract.

(D) The contractor will receive an annuity payment over some time period specified in the contract.

4. A tort is

(A) a civil wrong committed against another person

(B) a section of a legal contract

(C) a legal procedure in which complaints are heard in front of an arbitrator rather than a judge or jury

(D) the breach of a contract

5. If a contract does not include the boilerplate clause, "Time is of the essence," which of the following is true?

(A) It is difficult to recover losses for extra hours billed.

(B) Standard industry time guidelines apply.

(C) Damages for delay cannot be claimed.

(D) Workers need not be paid for downtime in the project.

6. Which statement is true regarding the legality and enforceability of contracts?

(A) For a contract to be enforceable, it must be in writing.

(B) A contract to perform illegal activity will still be enforced by a court.

(C) A contract must include a purchase order.

(D) Mutual agreement of all parties must be evident.

7. Which option best describes the contractual lines of privity between parties in a general construction contract?

(A) The consulting engineer will have a contractual obligation to the owner, but will not have a contractual obligation with the general contractor or the subcontractors.

(B) The consulting engineer will have a contractual obligation to the owner and the general contractor.

(C) The consulting engineer will have a contractual obligation to the owner, general contractor, and subcontractors.

(D) The consulting engineer will have a contractual obligation to the general contractor, but will not have a contractual obligation to the owner or subcontractors.

8. A contract has a value engineering clause that allows the parties to share in improvements that reduce cost. The contractor had originally planned to transport concrete on-site for a small pour with motorized wheelbarrows. On the day of the pour, however, a concrete pump is available and is used, substantially reducing the contractor's labor cost for the day. This is an example of

(A) value engineering whose benefit will be shared by both contractor and owner

(B) efficient methodology whose benefit is to the contractor only

(C) value engineering whose benefit is to the owner only

(D) cost reduction whose benefit will be shared by both contractor and laborers

9. In which of the following fee structures is a specific sum paid to the engineer for each day spent on the project?

(A) salary plus

(B) per-diem fee

(C) lump-sum fee

(D) cost plus fixed fee

10. What type of damages is paid when responsibility is proven but the injury is slight or insignificant?

(A) nominal

(B) liquidated

(C) compensatory

(D) exemplary

SOLUTIONS

1. In order to prove strict liability in tort, it must be shown that a product defect caused an injury. Negligence need not be proven, nor must the manufacturer know about the defect before release.

The answer is (D).

2. A material breach of the contract is a significant event that is grounds for cancelling the contract entirely. Typical triggering events include failure of the owner to make payments, the owner causing delays, the owner declaring bankruptcy, the contractor abandoning the job, or the contractor getting substantially off schedule.

The answer is (C).

3. Changes to a structure's performance, safety, appearance, or maintenance that benefit the owner in the long run will be covered by the value engineering clause of a contract. Normally, the contractor is able to share in cost savings in some manner by receiving a payment or credit to the contract.

The answer is (A).

4. A tort is a civil wrong committed against a person or his/her property which results in some form of damages. Torts are normally resolved through lawsuits.

The answer is (A).

5. This clause must be included in order to recover damages due to delay.

The answer is (C).

6. In order for a contract to be legally binding, it must

- be established for a legal purpose
- contain a mutual agreement by all parties
- have consideration, or an exchange of something of value (e.g., a service is provided in exchange for a fee)
- not obligate parties to perform illegal activity
- not be between parties that are mentally incompetent, minors, or do not otherwise have the power to enter into the contract

A contract does not need to use as its basis or include a purchase order to be enforceable. Oral contracts may be legally binding in some instances, depending on the circumstances and purpose of the contract. Oral contracts may be difficult to enforce, however, and should not be used for engineering and construction agreements.

The answer is (D).

7. With a general construction contract, a consulting engineer will be hired by the owner to develop the design and contract documents, as well as to assist in the preparation of the bid documents and provide contract administrative services during the construction phase. The contract documents produced by the engineer will form the basis of the owner's agreement with the contractor. Although the engineer will work closely with the contractor during the construction phase, and may work with subcontractors as well, the engineer will not have a contractual line of privity with either party.

The answer is (A).

8. The problem gives an example of efficient methodology, where the benefit is to the contractor only. It is not an example of value engineering, as the change affects the contractor, not the owner. Performance, safety, appearance, and maintenance are unaffected.

The answer is (B).

9. A specific fee is paid to the engineer for each day on the job in a per-diem fee structure.

The answer is (B).

10. Nominal damages are awarded for inconsequential injuries.

The answer is (A).

52 Ethics

PRACTICE PROBLEMS

1. An environmental engineer with five years of experience reads a story in the daily paper about a proposal being presented to the city council to construct a new sewage treatment plant near protected wetlands. Based on professional experience and the facts presented in the newspaper, the engineer suspects the plant would be extremely harmful to the local ecosystem. Which of the following would be an acceptable course of action?

(A) The engineer should contact appropriate agencies to get more data on the project before making a judgment.

(B) The engineer should write an article for the paper's editorial page urging the council not to pass the project.

(C) The engineer should circulate a petition through the community condemning the project, and present the petition to the council.

(D) The engineer should do nothing because he doesn't have enough experience in the industry to express a public opinion on the matter.

2. An engineer is consulting for a construction company that has been receiving bad publicity in the local papers about its waste-handling practices. Knowing that this criticism is based on public misperceptions and the paper's thirst for controversial stories, the engineer would like to write an article to be printed in the paper's editorial page. What statement best describes the engineer's ethical obligations?

(A) The engineer's relationship with the company makes it unethical for him to take any public action on its behalf.

(B) The engineer should request that a local representative of the engineering registration board review the data and write the article in order that an impartial point of view be presented.

(C) As long as the article is objective and truthful, and presents all relevant information including the engineer's professional credentials, ethical obligations have been satisfied.

(D) The article must be objective and truthful, present all relevant information including the engineer's professional credentials, and disclose all details of the engineer's affiliation with the company.

3. After making a presentation for an international project, an engineer is told by a foreign official that his company will be awarded the contract, but only if it hires the official's brother as an advisor to the project. The engineer sees this as a form of extortion and informs his boss. His boss tells him that, while it might be illegal in the United States, it is a customary and legal business practice in the foreign country. The boss impresses upon the engineer the importance of getting the project, but leaves the details up to the engineer. What should the engineer do?

(A) He should hire the official's brother, but insist that he perform some useful function for his salary.

(B) He should check with other companies doing business in the country in question, and if they routinely hire relatives of government officials to secure work, then he should do so too.

(C) He should withdraw his company from consideration for the project.

(D) He should inform the government official that his company will not hire the official's brother as a precondition for being awarded the contract, but invite the brother to submit an application for employment with the company.

4. If one is aware that a registered engineer willfully violates a state's rule of professional conduct, one should

(A) do nothing

(B) report the violation to the state's engineering registration board

(C) report the violation to the employer

(D) report the violation to the parties it affects

5. Which of the following is an ethics violation specifically included in the NCEES *Model Rules*?

(A) an engineering professor "moonlighting" as a private contractor

(B) an engineer investing money in the stock of the company for which he/she works

(C) a civil engineer with little electrical experience signing the plans for an electric generator

(D) none of the above

6. A senior licensed professional engineer with 30 years of experience in geotechnical engineering is placed in charge of a multidisciplinary design team consisting of a structural group, a geotechnical group, and an environmental group. In this role, the engineer is responsible for supervising and coordinating the efforts of the groups when working on large interconnected projects. In order to facilitate coordination, designs are prepared by the groups under the direct supervision of the group leader, and then they are submitted to her for review and approval. This arrangement is ethical as long as

(A) the engineer signs and seals each design segment only after being fully briefed by the appropriate group leader

(B) the engineer signs and seals only those design segments pertaining to geotechnical engineering

(C) each design segment is signed and sealed by the licensed group leader responsible for its preparation

(D) the engineer signs and seals each design segment only after it has been reviewed by an independent consulting engineer who specializes in the field in which it pertains

7. The National Society of Professional Engineers' (NSPE) Code of Ethics addresses competitive bidding. Which of the following is NOT stipulated?

(A) Engineers and their firms may refuse to bid competitively on engineering services.

(B) Clients are required to seek competitive bids for design services.

(C) Federal laws governing procedures for procuring engineering services (e.g., competitive bidding) remain in full force.

(D) Engineers and their societies may actively lobby for legislation that would prohibit competitive bidding for design services.

8. You are a city engineer in charge of receiving bids on behalf of the city council. A contractor's bid arrives with two tickets to a professional football game. The bid is the lowest received. What should you do?

(A) Return the tickets and accept the bid.

(B) Return the tickets and reject the bid.

(C) Discard the tickets and accept the bid.

(D) Discard the tickets and reject the bid.

9. A relatively new engineering firm is considering running an advertisement for their services in the local newspaper. An ad agency has supplied them with four concepts. Of the four ad concepts, which one(s) would be acceptable from the standpoint of professional ethics?

I. an advertisement contrasting their successes over the past year with their nearest competitors' failures

II. an advertisement offering a free television to anyone who hires them for work valued at over $10,000

III. an advertisement offering to beat the price of any other engineering firm for the same services

IV. an advertisement that tastefully depicts their logo against the backdrop of the Golden Gate Bridge

(A) I and III

(B) I, III, and IV

(C) II, III, and IV

(D) neither I, II, III, nor IV

10. Complete the sentence: "A professional engineer who took the licensing examination in mechanical engineering

(A) may not design in electrical engineering."

(B) may design in electrical engineering if she feels competent."

(C) may design in electrical engineering if she feels competent and the electrical portion of the design is insignificant and incidental to the overall job."

(D) may design in electrical engineering if another engineer checks the electrical engineering work."

11. An engineering firm is hired by a developer to prepare plans for a shopping mall. Prior to the final bid date, several contractors who have received bid documents and plans contact the engineering firm with requests for information relating to the project. What can the engineering firm do?

(A) The firm can supply requested information to the contractors as long as it does so fairly and evenly. It cannot favor or discriminate against any contractor.

(B) The firm should supply information to only those contractors that it feels could safely and economically perform the construction services.

(C) The firm cannot reveal facts, data, or information relating to the project that might prejudice a contractor against submitting a bid on the project.

(D) The firm cannot reveal facts, data, or information relating to the project without the consent of the client as authorized or required by law.

SOLUTIONS

1. The engineer certainly has more experience and knowledge in the field than the general public or even the council members who will have to vote on the issue. Therefore, the engineer is qualified to express his opinion if he wishes to do so. Before the engineer takes any public position, however, the engineer is obligated to make sure that all the available information has been collected.

The answer is (A).

2. It is ethical for the engineer to issue a public statement concerning a company he works for, provided he makes that relationship clear and provided the statement is truthful and objective.

The answer is (D).

3. Hiring the official's brother as a precondition for being awarded the contract is a form of extortion. Depending on the circumstances, however, it may be legal to do so according to U.S. law. (The Foreign Corrupt Practices Act of 1977 allows American companies to pay extortion in some cases.) This practice, however, is not approved by the NCEES *Model Rules*:

> Registrants shall not offer, give, solicit, or receive, either directly or indirectly, any commission or gift, or other valuable consideration in order to secure work.

The answer is (D).

4. A violation should be reported to the organization that has promulgated the rule.

The answer is (B).

5. The NCEES *Model Rules* specifically states that registrants may not perform work beyond their level of expertise. The other two examples may be unethical under some circumstances, but are not specifically forbidden by the NCEES code.

The answer is (C).

6. According to the NCEES *Model Rules*,

> Licensees may accept assignments for coordination of an entire project, provided that each design segment is signed and sealed by the registrant responsible for preparation of that design segment.

The answer is (C).

7. Clients are not required to seek competitive bids. In fact, many engineering societies discourage the use of bidding to procure design services because it is believed that competitive bidding results in lower-quality construction.

The answer is (B).

8. Registrants should not accept gifts from parties expecting special consideration, so the tickets cannot be kept. They also should not be merely discarded, for several reasons. Inasmuch as the motive of the contractor is not known with certainty, in the absence of other bidding rules, the bid may be accepted.

The answer is (A).

9. None of the ads is acceptable from the standpoint of professional ethics. Concepts I and II are explicitly prohibited by the NCEES *Model Rules*. Concept III demeans the profession of engineering by placing the emphasis on price as opposed to the quality of services. Concept IV is a misrepresentation; the picture of the Golden Gate Bridge in the background might lead some potential clients to believe that the engineering firm in question had some role in the design or construction of that project.

The answer is (D).

10. Although the laws vary from state to state, engineers are usually licensed generically. Engineers are licensed as "professional engineers." The scope of their work is limited only by their competence. In the states where the license is in a particular engineering discipline, an engineer may "touch upon" another discipline when the work is insignificant and/or incidental.

The answer is (C).

11. It is normal for engineers and architects to clarify the bid documents. However, some information may be proprietary to the developer. The engineering firm should only reveal information that has already been publicly disseminated or approved for release with the consent of the client.

The answer is (D).

53 Licensure

PRACTICE PROBLEMS

There are no problems in this book covering the subject of licensure.